Wissenschaftliche Reihe Fahrzeugtechnik Universität Stuttgart

Das Institut für Fahrzeugtechnik Stuttgart (IFS) an der Universität Stuttgart erforscht, entwickelt, appliziert und erprobt, in enger Zusammenarbeit mit der Industrie, Elemente bzw. Technologien aus dem Bereich moderner Fahrzeugkonzepte. Das Institut gliedert sich in die drei Bereiche Kraftfahrwesen, Fahrzeugantriebe und Kraftfahrzeug-Mechatronik. Aufgabe dieser Bereiche ist die Ausarbeitung des Themengebietes im Prüfstandsbetrieb, in Theorie und Simulation. Schwerpunkte des Kraftfahrwesens sind hierbei die Aerodynamik, Akustik (NVH), Fahrdynamik und Fahrermodellierung, Leichtbau, Sicherheit, Kraftübertragung sowie Energie und Thermomanagement – auch in Verbindung mit hybriden und batterieelektrischen Fahrzeugkonzepten. Der Bereich Fahrzeugantriebe widmet sich den Themen Brennverfahrensentwicklung einschließlich Regelungs- und Steuerungskonzeptionen bei zugleich minimierten Emissionen, komplexe Abgasnachbehandlung, Aufladesysteme und -strategien, Hybridsysteme und Betriebsstrategien sowie mechanisch-akustischen Fragestellungen. Themen der Kraftfahrzeug-Mechatronik sind die Antriebsstrangregelung/Hybride, Elektromobilität, Bordnetz und Energiemanagement, Funktions- und Softwareentwicklung sowie Test und Diagnose. Die Erfüllung dieser Aufgaben wird prüfstandsseitig neben vielem anderen unterstützt durch 19 Motorenprüfstände, zwei Rollenprüfstände, einen 1:1-Fahrsimulator, einen Antriebsstrangprüfstand, einen Thermowindkanal sowie einen 1:1-Aeroakustikwindkanal. Die wissenschaftliche Reihe „Fahrzeugtechnik Universität Stuttgart" präsentiert über die am Institut entstandenen Promotionen die hervorragenden Arbeitsergebnisse der Forschungstätigkeiten am IFS.

Reihe herausgegeben von

Prof. Dr.-Ing. Michael Bargende
Lehrstuhl Fahrzeugantriebe
Institut für Fahrzeugtechnik Stuttgart
Universität Stuttgart
Stuttgart, Deutschland

Prof. Dr.-Ing. Hans-Christian Reuss
Lehrstuhl Kraftfahrzeugmechatronik
Institut für Fahrzeugtechnik Stuttgart
Universität Stuttgart
Stuttgart, Deutschland

Prof. Dr.-Ing. Jochen Wiedemann
Lehrstuhl Kraftfahrwesen
Institut für Fahrzeugtechnik Stuttgart
Universität Stuttgart
Stuttgart, Deutschland

Marvin Sascha Wahl

Emission Reduction with an Alternative Diesel Combustion Process

Springer Vieweg

Marvin Sascha Wahl
Institute of Automotive Engineering
(IFS), Chair in Automotive Powertrains
University of Stuttgart
Stuttgart, Germany

Zugl.: Dissertation Universität Stuttgart, 2023
D93

ISSN 2567-0042 ISSN 2567-0352 (electronic)
Wissenschaftliche Reihe Fahrzeugtechnik Universität Stuttgart
ISBN 978-3-658-42093-2 ISBN 978-3-658-42094-9 (eBook)
https://doi.org/10.1007/978-3-658-42094-9

This Springer Vieweg imprint is published by the registered company Springer Fachmedien Wiesbaden GmbH, part of Springer Nature.
The registered company address is: Abraham-Lincoln-Str. 46, 65189 Wiesbaden, Germany

Preface

Die vorliegende Arbeit entstand während meiner Tätigkeit als wissenschaftlicher Mitarbeiter am Institut für Fahrzeugtechnik Stuttgart (IFS) der Universität Stuttgart unter der Leitung von Herrn Prof. Dr.-Ing. M. Bargende und Herrn Prof. Dr.-Ing. A. Casal Kulzer. Ihnen gilt mein besonderer Dank für die wissenschaftliche und persönliche Betreuung während dieser Arbeit.

Herrn Prof. Dr.-Ing. Peter Eilts danke ich herzlich für das entgegengebrachte Interesse an der Arbeit und für die Übernahme des Koreferates.

Allen Kolleginnen und Kollegen des IFS sowie des FKFS danke ich für die Unterstützung und die unvergessliche Zeit während meiner Tätigkeit am Institut. Des Weiteren danke ich allen studentischen Hilfskräften für die Projektunterstützung. Mein besonderer Dank gilt Herrn Dipl.-Ing. Hans-Jürgen Berner. Die facettenreichen Gespräche zu später Stunde sowie seine Hilfsbereitschaft werden mir immer im Gedächtnis bleiben.

Der FVV e.V. gilt mein Dank für die Finanzierung des Projekts „Premixed Diesel". Die Ergebnisse dieser Dissertation sind in diesem Rahmen entstanden und im zugehörigen Abschlussbericht bereits veröffentlicht. Ebenso gilt mein persönlicher Dank dem Obmann Herrn Dr. S. Schneider der MAHLE International GmbH für seine Unterstützung.

Einen essenziellen Faktor während meiner Tätigkeit stellte der Rückhalt meiner Familie und Freunde dar. Ich möchte mich hierzu besonders bei meiner Schwester, meiner Mutter und meinem Vater für ihr Verständnis und ihre Hilfsbereitschaft bedanken. Auch möchte ich Manuel Ress für die Freundschaft und die vielen Stunden in der Werkstatt danken.

HK

Gerabronn — Marvin Wahl

Contents

Figures

Tables

Abbreviations

AFR	Air-Fuel Ratio (A/F)
AMB	Ambient
ANR	Analyzer Rack
BSFC	Brake Specific Fuel Consumption
C_3H_8	Propane
CAI	Controlled Auto Ignition
CDM	Crank Degree Marks
cf	Confer (Latin, meaning "Compare")
CLD	Chemiluminescence Detector
CN	Cetane Number
CO	Carbon Monoxide
CO_2	Carbon Dioxide
COV	Coefficient of Variation
CVS	Constant Volume Sampling
DCCS	Dilution Controlled Combustion System
DCPCI	Distribution Controlled Partially Premixed Compression Ig-
DI	Direct Injection
DME	Dimethyl Ether
DOC	Diesel Oxidation Catalyst
ECU	Electronic (Engine) Control Unit
EGR	Exhaust Gas Recirculation
EKAS	Intake Port Closing with Swirl Flap (Einlasskanal Abschal-
EOC	End of Combustion
EPA	Environmental Protection Agency
Et Al.	Et Alii (Latin, meaning "And Other People")

EVC	Exhaust Valve Closes
EVO	Exhaust Valve Opens
FAME	Fatty Acid Methyl Ester
FID	Flame Ionization Detector
FSN (BSN)	Filter Smoke Number (also known as Bosch Smoke Num-
HCCI	Homogeneous Charge Compression Ignition
HCLI	Homogeneous Charge Late Ignition
HPLI	Highly Premixed Late Injection
HTHR	High Temperature Heat Release
IC	Interface Controller
ID	Ignition Delay
IDI	In-Direct Injection
IMEP	Indicated Mean Effective Pressure (IMEP or $p_{m,i}$)
INCA	ETAS GmbH Software
IQA	Injection Quantity Adjustment
ISFC	Indicated Specific Fuel Consumption
IVC	Intake Valve Closes
IVO	Intake Valve Opens
LNT	Lean NO_X Trap
LTHR	Low Temperature Heat Release
MCU	Main Control Unit
MPA	Magnetic Pressure (Oxygen) Analyzer
NDIR	Non-Dispersive Infrared
NO	Nitric Oxide
NO_2	Nitrogen Dioxide
NO_X	Nitrogen Oxides
NTC	Negative Temperature Coefficient
O_2	Oxygen

OVN	Oven
PAH	Polycyclic Aromatic Hydrocarbon
PCCI	Premixed Charge Compression Ignition
PFI	Port Fuel Injection
PHEV	Plug-In Hybrid Electric Vehicle
PM	Particulate Matter
ppm	Parts per Million
PSU	Power Supply Unit
RME	Rapeseed (Oil) Methyl Ester
ROHR	Rate of Heat Release
SCP	Signal Conditioning Platform
SCR	Selective Catalytic Reduction
SHS	Sample Handling System
SI	Spark Ignition
SMD	Sauter Mean Diameter
SO_2	Sulfur Dioxide
SO_3	Sulfur Trioxide
SOC	Start of Combustion
SOF	Soluble Organic Fraction
SOI	Start of Injection
SVS	Solenoid Valve Selector
TDC	Top Dead Center
THC	Total Hydrocarbon
TRG	Trigger Signal
TWC	Three-Way Catalyst

Symbols

α	[°]	Crank Angle
α_k	[-]	Alpha k
A_P	[cm^2]	Piston Cross Sectional Area
A_{is}	[cm^2]	Isentropic Area
H_l	[J]	Enthalpy (Leakage)
H_u	$\left[\frac{MJ}{kg}\right]$	Lower Heating Value
Q_b	[J]	Heat
Q_w	[J]	Heat (Wall)
V_c	[m^3]	Clearance Volume
V_d	[m^3]	Displacement Volume
c_p	$\left[\frac{J}{kgK}\right]$	Specific Heat Capacity (Constant Pressure)
c_v	$\left[\frac{J}{kgK}\right]$	Specific Heat Capacity (Constant Volume)
d_P	[cm^2]	Piston Cross Section Diameter
m_c	[kg]	Cylinder Mass
m_e	[kg]	Exhaust Mass
m_f	[kg]	Fuel Mass
m_i	[kg]	Inlet Mass
m_l	[kg]	Leakage Mass
w_φ	$\left[\frac{MJ}{kg}\right]$	Burn Rate (Wiebe)
η_e	[-]	Conversion Efficiency
λ_l	[-]	Connecting Rod Ratio
φ_d	[°]	Combustion Duration (Wiebe)

°CAaBDC	[°]	Degree Crank Angle after Bottom Dead Center
°CAaTDC	[°]	Degree Crank Angle after Top Dead Center
°CAbBDC	[°]	Degree Crank Angle before Bottom Dead Center
°CAbTDC	[°]	Degree Crank Angle before Top Dead Center
p	[bar]	Pressure
R	$\left[\frac{J}{kgK}\right]$	Mass Specific Gas Constant
T	[K]	Temperature
v	$\left[\frac{m^3}{kg}\right]$	Specific Volume
V	[m^3]	Volume
U	[J]	Internal Energy
W	[J]	Work
m	[-]	Wiebe Form Factor
ε	[-]	Compression Ratio
φ	[°]	Crank Angle

Kurzfassung

Die Entwicklung von Diesel- sowie Ottomotoren schreitet seit einigen Jahrzenten schnell voran. Die fortschreitenden Verschärfungen der Emissionsgesetzgebung von Klein- bis hin zu Großmotoren waren dabei ebenso die Ziele der Entwicklung, wie auch die Verbesserung hinsichtlich der Eigenschaften wie Kraftstoffverbrauch, Fahrbarkeit, Preis und Gewicht. Dies führte zu deutlicher Reduktion der Rohemissionen durch Optimierung der Verbrennung sowie deren weitere Nachbehandlung mit Systemen der Abgasnachbehandlung. Der Umstieg auf neuartige Einspritzsysteme, durch welche die Motorrohemissionen gesenkt und gleichzeitig das Ansprechverhalten verbessert werden konnten, stand ebenso im Fokus wie die externe Aufladung, das Downsizing und verbesserte Eigenschaften für transientes Motor- sowie Kaltstartverhalten. Verschiedene Entwicklungen wurden miteinander kombiniert, um die Motoren für gültige gesetzliche Vorschriften und den Einsatzzweck beim Anwender zu optimieren. Im Bereich der PKW-Motoren liegt der Fokus vor allem auf der Fahrbarkeit und dem Laufverhalten im transienten, hochdynamischen Motorbetrieb. Dabei ist anhand von Messungen der „Real Driving Emissions" (RDE) die Einhaltung der gültigen Abgasemissionsvorschriften nachzuweisen. Eine weitere Verschärfung der derzeitig gültigen Vorschriften in Europa wird voraussichtlich als EU7 veröffentlicht und stellt einen weiteren Entwicklungsschritt aller PKW-Motoren dar.

Um den gültigen Vorschriften nachzukommen sind in der Vergangenheit einige Optimierungen umgesetzt worden. Eine weitere Möglichkeit zur innermotorischen Reduktion der Rohemissionen leitet sich aus dem Grenzfall der homogenen Gemischaufbereitung ab. Die Verdampfungseigenschaften von Otto-Kraftstoff ermöglicht dessen Einbringung anhand eines Vergasers oder durch Einspritzdüsen in den Luftpfad des Motors. Eine ähnliche Einbringung von Dieselkraftstoff ist durch dessen Kraftstoffeigenschaften nicht möglich. Untersuchungen zur Einbringung von Dieselkraftstoff und dieselähnlichen Kraftstoffen vor den Einlassventilen wurden jedoch in vorangegangenen Projekten durchgeführt. Die Verdampfung, Vermischung sowie die Selbstzündung wurden dabei für verschiedene Parameter untersucht. Es konnte ein

Atomizer eingesetzt werden, durch welchen der Kraftstoff aufgewärmt und fein zerstäubt in das Saugrohr eines Einzylinder-Dieselmotors eingebracht wurde. Diese Einspritzstrategie der homogenen Kompressionszündung (HCCI – *Homogeneous Charge Compression Ignition*) wurde für grundlegende Untersuchungen bezüglich des Zündverzuges und der Homogenisierung des Dieselkraftstoffes vor dem Start der Verbrennung (SOC – *Start of Combustion*) angewandt. Des Weiteren wurden Experimente zur Bestimmung der Cetanzahl und der zugehörigen Zündverzugszeit verschiedener Kraftstoffmischungen in einem BASF-Versuchsmotor durchgeführt. Um diese Erkenntnisse in realitätsnahe Projekte zu übertragen, sollte ein Ansatz zur Anwendung einer Einspritzstrategie in modernen Motoren mittels Common-Rail Einspritzung gefunden werden. Dabei war einerseits der Drehzahl- und Lastbereich des Motors für einen möglichen Serieneinsatz entscheidend, andererseits sollten die Vorteile der Homogenisierung vor Verbrennungsbeginn berücksichtigt werden. Eine Möglichkeit bietet die teilvorgemischte Dieselverbrennung (PCCI – *Premixed Charge Compression Ignition*). Dabei sollen die Vorteile der homogenen Verbrennung mit den Vorteilen der konventionellen Dieselverbrennung kombiniert werden.

In der vorliegenden Arbeit wurde ein Einzylindermotor optimiert, um die jeweiligen Vorteile von PCCI und des konventionell betriebenen Dieselmotors im gleichen Brennraum zu untersuchen. Das geometrische Verdichtungsverhältnis wurde von 18:1 auf 15,34:1 reduziert, um die Selbstzündung des voreingespritzten Kraftstoffes durch die Beeinflussung des Zustandes im Zylinder zu verzögern. Mit den verwendeten mechanischen Motorkomponenten – Injektor, Kolben, Zylinderkopf und Pleuel – war somit weiterhin der Betrieb mit hoher Last und Drehzahl möglich. Dieses Projekt beschäftigte sich vor allem mit der Einspritzstrategie der teilvorgemischten Dieselverbrennung. Diese konnte bis 2000 min^{-1} und 10 bar indizierten Mitteldruck beziehungsweise 2500 min^{-1} und 7 bar indizierten Mitteldruck eingesetzt werden. Dazu war ein Anteil von 30–40 % externer Abgasrückführrate (AGR) zur Verschiebung des Verbrennungsschwerpunktes notwendig. Dieser Anteil der AGR steuerte zugleich die Umsetzungsrate und dadurch den Druckanstiegsgradienten der Verbrennung. Der Luftpfad ist – entgegen dem Kraftstoffpfad – jedoch sehr träge, weshalb ein dynamischer Kennfeldbetrieb im Grenzbereich von PCCI nur schwer realisierbar ist. Transiente Fahrprofile können somit nicht von der

Verbrennungskraftmaschine realisiert werden, sondern bedürfen bei Einsatz von PCCI einer hybriden Antriebsstrangkonfiguration zur Lastpunktverschiebung.

Des Weiteren erfordert die Verbrennungscharakteristik die Einführung von Sicherheitsfaktoren, durch welche die Motormechanik vor Schädigung geschützt wird. Diese bewirken unter anderem die Begrenzung des maximalen Zylinderdruckes sowie des maximalen Druckanstiegsgradienten. Als Referenz dient hierzu der Betriebspunkt aus Kapitel 4.5 bei 2000 min^{-1} und 10 bar indiziertem Mitteldruck. Wie in Abbildung 4.15 ersichtlich, ergibt sich dafür bei konventioneller Dieselverbrennung ein Spitzendruck von 95 bar und ein maximaler Druckanstieg pro Grad Kurbelwinkel von 8,7 bar. Für die teilvorgemischte Dieselverbrennung ergibt sich für denselben Betriebspunkt ein Spitzendruck von 159 bar bei einem Druckanstieg 12,22 bar. Dieser Spitzendruck, einhergehend mit dem Druckanstieg, ist für Serienanwendungen aufgrund des Motorgeräusches nicht anwendbar. Da die Motormechanik bis zu einem maximalen Druck von 175 bar ausgelegt ist, werden diese Messpunkte jedoch als valide Grenzmessungen einbezogen. Die Anregung der Motorstruktur aus der Verbrennungscharakteristik ist aus akustischen Gründen somit inakzeptabel, jedoch ist die mechanische Beanspruchung aufgrund des Druckniveaus tolerierbar.

Die charakteristische Homogenisierung der alternativen Einspritzstrategie trägt durch die Auflösung lokaler Inhomogenitäten zur Reduzierung von sowohl Partikel- als auch Stickoxidbildung bei. Die folgende Abbildung, welche aus Abbildung 1.8 abgeleitet ist, zeigt dabei schematisch die Bereiche der Emissionsbildung. Die Bereiche der konventionellen Dieselverbrennung sowie verschiedener alternativer Brennverfahren sind zur Verdeutlichung aufgeführt. Die alternativen Brennverfahren sind in Abbildung 1.5 genauer definiert und gegenübergestellt. Die Motormessungen wurden dabei für verschiedene Betriebspunkte im Kennfeld durchgeführt. Als Grundlage der Überlegungen zur Reduktion von Partikel- und Stickoxidemissionen konnte die nachfolgende Abbildung herangezogen werden. Diese Abbildung zeigt die Bereiche auf, in denen diese beiden Abgasbestandteile in Abhängigkeit von der lokalen Flammentemperatur und dem Luft-/Kraftstoffverhältnis gebildet werden. Für die konventionelle Dieselverbrennung ergibt sich daraus der Kompromiss zwischen Partikel- und Stickoxidbildung. Die Bildung von Partikeln tritt dabei maßgeblich in lokal unterstöchiometrischen, also kraftstoffreichen Zonen auf. Die Stickoxidbildung dagegen in leicht überstöchiometrischen, kraftstoffarmen Bereichen. Einzelne Phänomene und Bildungsmechanismen werden in der vorliegenden Arbeit weiter erläutert.

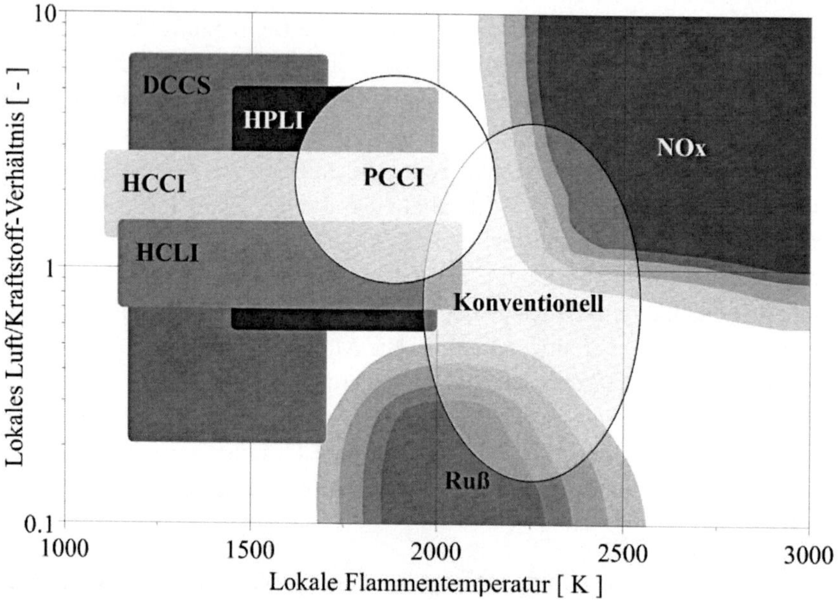

Der primäre- und sekundäre Strahlaufbruch sowie die Verdampfung des Dieselkraftstoffes reduzieren die Bildungsmechanismen von aromatischen Ringstrukturen und polyzyklischen aromatischen Kohlenwasserstoffen. Diese sind ausschlaggebend für das Wachstum weiterer Ringstrukturen und folglich von Partikelemissionen. Die zeitliche und örtliche Unabhängigkeit zwischen Einspritzende und Brennbeginn bei alternativen Brennverfahren fördert zusätzlich die Durchmischung mit der vorhandenen Ladung, wodurch zwar das Luft-/Kraftstoffverhältnis im Bereich der Stickoxidbildung liegt, jedoch die nach Zeldovich beschriebene Bildungstemperatur zu gering ist.

Für das hier untersuchte teilvorgemischte Dieselbrennverfahren (PCCI) sind gleichzeitig folgende Bedingungen zu erfüllen:

- Das Ausbleiben diffusiver Verbrennung.
- Einspritzende möglichst vor Verbrennungsbeginn.
- Reduktion der Brennrate und ausmagern lokaler Bereiche.
- Möglichst großer Anteil teilhomogener Verbrennung.

Dafür konnten am Einzylinder-Motorenprüfstand verschiedene Parametervariationen hinsichtlich Motorlast und -drehzahl, externe Abgasrückführrate, Ansauglufttemperatur, Druckverhältnis (p_2/p_3), Kolbenbodenkühlung, Einspritzstrategie und Kraftstoffdruck untersucht werden. Zum Vergleich der konventionellen und teilvorgemischten Einspritzstrategie wurde ein Last- und Drehzahlschnitt durchgeführt. Die Motorrohemissionen von PCCI und konventioneller Einspritzung wurden verglichen und die jeweiligen Vor- und Nachteile herausgearbeitet. Die Ergebnisse der Untersuchungen können wie folgt zusammengefasst werden:

- Die frühe Voreinspritzung von Kraftstoff hat den Nachteil der Oberflächenbenetzung und der frühen Selbstzündung. Dies führt zu erhöhten CO und HC-Emissionen sowie zu einem erhöhten spezifischen Kraftstoffverbrauch verglichen mit konventioneller Einspritzstrategie.
- Ein Vorteil stellt die gleichzeitige Reduktion von Stickoxid- und Partikelemissionen dar. Innermotorisch konnten dafür zugleich die Verbrennungstemperatur als auch lokal fette Bereiche reduziert werden, wodurch die Bildungsmechanismen beider Schadstoffe teilweise umgangen werden konnten.
- Die Menge des voreingespritzten Kraftstoffs ist durch den Wandauftrag begrenzt, jedoch hat die Verteilung (Kraftstoffmenge) der Vor- und Haupteinspritzung eine große Bedeutung bei der Reduktion der Emissionen. Nach den Voreinspritzungen zeigte ein globales Luft-/Kraftstoffverhältnis zwischen 2,9 und 4,6 das beste Reduktionspotential. Die externe Abgasrückführrate war dabei zwischen 30 und 40 %.

Der direkte Vergleich zur konventionellen Einspritzstrategie zeigt verschiedene Merkmale auf, wobei PCCI…

- …einen um bis zu 7 % höheren spezifischen Verbrauch…
- …geringere (an der Nachweisgrenze) Rußemissionen…
- …geringere (an der Nachweisgrenze) Stickoxidemissionen…
- …höhere THC und CO-Emissionen…
- …höhere Spitzendrücke und Druckanstiegsgradienten…
- …eine geringere Drehzahl- und Lastgrenze…
- …ein erhöhtes Geräuschniveau…

aufweist. Hierbei müssen die gemessenen Betriebspunkte genauer betrachtet werden. Auch wurden für die konventionellen Betriebspunkte keine Variationen zur Konditionierung der Abgasnachbehandlung berücksichtigt. Somit sind weder Fettsprünge noch Heizstrategien betrachtet worden, weshalb der Nachteil von PCCI teilweise überproportional gewichtet ist.

Für weitere Untersuchungen von PCCI würde sich eine optimierte Kolbengeometrie sowie eine angepasste Düsengeometrie des Injektors anbieten. Daraus könnte sich weiteres Potential durch eine bessere Durchmischung und weniger Wandauftrag erschließen, wobei als möglicher Einspritzzeitpunkt der Voreinspritzungen 60 bis 20 Grad vor dem oberen Totpunkt in Betracht gezogen werden sollte. Eine deutlich erhöhte Turbulenz nahe dem oberen Totpunkt könnte die Sprayverdampfung weiter fördern und somit die Tropfengröße und Tropfengrößenverteilung positiv beeinflussen. Der druckbildende Anteil sowie der Schwerpunkt der Verbrennung könnte aufgrund der kürzeren Zeitskala für den Zündverzug besser geregelt werden. Teilweise erschließen sich diese Aussagen aus dem dargestellten Schaubild in Kapitel 4.3.1 und konnten während dem Prüfstandsbetrieb mit positivem Einfluss auf die Verbrennung identifiziert werden.

Aus den Messdaten wurde zusätzlich ein nulldimensionales Zündverzugsmodell entwickelt. Dem Verfasser ist keine Publikation bekannt, die weder eine so große Datenbasis mit unterschiedlichen Variationsparametern noch PCCI mit Mehrfacheinspritzung betrachtet. Der Entwicklungsschwerpunkt war hierbei die Nieder- und Hochtemperaturzündung des voreingespritzten Kraftstoffes, da der SOC aufgrund von langem Zündverzug bisher als nicht vorhersagefähig galt. Auf der Grundlage des ursprünglichen Arrhenius-Ansatzes wurden in der Vergangenheit Zündverzugsmodelle an den Dieselmotor angepasst, wobei der daraus hervorgegangene dreistufige Arrhenius-Ansatz für die Vorhersage der Nieder- und Hochtemperaturzündung in Abschnitt 5 weiterentwickelt und validiert wurde. Dieser basiert auf einer Publikation von Weisser, bei welcher zum ersten Mal der negative Temperaturkoeffizient zwischen der Nieder- und Hochtemperaturzündung berücksichtigt wurde. Die Anwendung eines Zündintegrals macht den dreistufigen Arrhenius Ansatz in seiner abgewandelten Form dann zur Vorhersage anwendbar. Eine grundlegende Frage bei Zündverzugsmodellen mit früherer (Teil)Homogenisierung stellt sich bezüglich des Rechenbeginns. Dieser kann entweder durch einen festen Kurbelwinkel – wie dies bei HCCI durch sehr frühe Einbringung des Kraftstoffes sinnvoll ist – oder auf den Start der Einspritzung definiert werden. Im vorliegenden Fall hatte sich der Beginn der ersten Voreinspritzung als sinnvoll erwiesen. Daraus ergeben sich deutliche Unterschiede der Zündverzugszeit, da sehr frühe (um 100 Grad vor OT) bis hin zu sehr späten (bis 20 Grad vor OT) Einspritzungen über das gleiche empirische Modell abgedeckt werden sollten. Die Zündverzugszeit über den Start der Einspritzung für die ausgewerteten Messpunkte ist aus Abbildung 5.1 zu entnehmen. Es zeigt sich bei späten Einspritzungen und kurzen Zündverzugszeiten (rechts unten im Bild) eine lineare Abhängigkeit von Einspritzzeitpunkt und Zündverzugszeit (mit blauer und roter Linie gekennzeichnet). Bei früheren Einspritzzeitpunkten löst sich die Abhängigkeit teilweise auf und es ergibt sich für die Nieder- und Hochtemperaturzündung die jeweils dargestellte schraffiertes Fläche in C-Form.

Die hier vorgestellten Ergebnisse der teilvorgemischten Dieselstrategie mit Mehrfacheinspritzung für seriennahe Motorkonfiguration zeigen die Möglichkeiten zur gleichzeitigen Reduktion von Partikel- und Stickoxidemissionen. Die konventionelle Dieselverbrennung hat dagegen weiterhin Vorteile im spezifischen Verbrauch und in den Geräuschemissionen. Die im Projekt erzielte Last- und Drehzahlerhöhung mit einem PCCI-Verfahren erlaubt eine Neubewertung der alternativen Brennverfahren, da dadurch der mittlere Lastbereich umfangreich abgedeckt werden kann. Ein transienter Betrieb mit den dargestellten Bestpunkten kann in Kombination mit einem hybriden Antriebsstrang umgesetzt werden. Die Lastpunktverschiebung im Kennfeld würde durch den elektrischen Anteil vorgenommen werden. Daraus würde sich ein „Umschalten[1]“ zwischen PCCI und konventionellem Betrieb ergeben. Dies müsste durch weitere Messungen untersucht und validiert werden.

Die in dieser Arbeit genutzten Messdaten wurden im Projekt „Premixed Diesel“ der FVV e.V. unter der Projektnummer 1352 generiert. Teilergebnisse wurden bereits im Abschlussbericht des Projektes veröffentlicht.

[1] Als „Umschalten“ ist hier der Betriebspunktwechsel gemeint. Dieser wird maßgeblich über den Wechsel von niedriger zu hoher AGR Rate bestimmt. Der Übergangsbereich kann weder als konventioneller noch als teilvorgemischter Betriebsbereich angesehen werden.

Abstract

Developments for diesel and gasoline engines have been progressing rapidly for several decades. From small to large engines, one of the major drivers of all development work has been the emission behavior of the engines. This includes approaches to reduce exhaust emissions from combustion and the possibilities of exhaust gas aftertreatment, which have been developed and implemented to the point where they were ready for series production. The adoption of new types of injection systems, which reduce engine emissions and at the same time improve engine response, has also been a focus, as were external turbocharging, downsizing, and transient engine as cold start behavior. Various developments were combined to optimize the engines for their intended application by the user and meet regulatory requirements. On the user side, the focus is on drivability in transient engine operation. For this highly dynamic application, compliance with exhaust emission regulations has become required with real driving emissions (RDE).

In previous projects, the diesel self-ignition behavior for alternative injection patterns was investigated. Therefore, different fuel mixtures were used in a BASF test engine in which the cetane number (CN) and the associated ignition delay were determined. In addition, a single-cylinder diesel engine was operated with a homogeneous charge compression ignition (HCCI) strategy. In the following steps, the results on ignition delay and homogenization of diesel before the start of combustion (SOC) were to be transferred into a realistic approach for use in state-of-the-art engines. On the one hand, a wide engine speed and load range were crucial for serial use. On the other hand, the advantages of concurrent low soot and NO_X engine-out emissions with partially premixed injection strategies as HCCI or premixed charge compression ignition (PCCI) in contrast to a soot and NO_X trade-off of conventional diesel combustion seems to be a sustainable path for upcoming diesel engine generations.

The mechanical parts in a state-of-the-art engine were modified to simultaneously take advantage of PCCI and conventional diesel strategies within the same combustion chamber. For this objective, the geometrical compression ratio was adjusted from 18:1 to 15.34:1. The retardation of the auto-ignition of pre-injected fuel for partly homogeneous conditions was achieved by influencing in-cylinder ratios and conditions. Nevertheless, high loads and engine speeds were still possible with the same mechanical parts applied in the conventional diesel engine. These include the injector, piston, cylinder head, and conrod. Initially, the focus was on engine measurements with PCCI and multiple injections. These results can be summarized as follows:

- Fuel pre-injection has the disadvantage of wall wetting and early auto-ignition.
- The partially premixed air-fuel ratio before SOC reduces locally rich areas (soot).
- Ignition of the early injected fuel does not lead to increased nitric oxide formation since the ignition locations are already very lean at SOC, and the combustion temperature remains low.
- Multiple pre-injections reduce CO and HC emissions compared to a single pre-injection with the same amount of fuel injected. Therefore, the amount of pre-injected fuel is limited, but a distribution of pre- and main-injected fuel leads to ultra-low soot and nitrogen oxide raw emissions.

Different variations were performed concerning engine load and speed, EGR rate, intake air temperature, pressure ratio (p_2/p_3), piston cooling, injection strategy, and fuel pressure. Further engine measurements with sweeps in load and speed were performed with a conventional diesel injection strategy to compare with PCCI. The engine-out emissions, performance, noise, and other parameters of PCCI and conventional injection strategy are compared in this work.

Additionally, a zero-dimensional ignition delay (ID) model was developed, focusing on the ignition delay of low and high-temperature combustion of pre-injected fuels. Based on the initial Arrhenius approach, a three-stage Arrhenius was recently developed and published by other researchers. The extension and adaption of the published approaches with respect to the variabilities described above were carried out and are described in detail in section 5.

The measurement data used in this work were obtained in the "Premixed Diesel" project of FVV e.V. under project number 1352. Some results have been published in the conference proceedings of the FVV e.V., the final report, at the Stuttgart Symposium [1], and at Thiesel2022 [2] in Valencia.

1 Introduction and State of Research

Retrospect to the year 1892: Rudolf Diesel – starting his career in Paris to develop a machine producing glaze ice – enrolled a patent with the number DRP 67 207 to improve the efficiency of internal combustion engines. This application describes the theory of diffusive combustion in a reciprocal engine. The compression stroke compresses the "pure" air or steam to the maximum cycle pressure. During this process, the temperature in the cylinder rises and exceeds the ignition temperature of the fuel that has not yet been injected. At this point, the gradual injection of fine-distributed fuel into the combustion chamber results in auto-ignition after a certain period, the so-called ignition delay time. Continuous injection and combustion further expand the working gas during the cycle's expansion phase. Thereby, the volume in the combustion chamber increases, and concurrently the in-cylinder pressure decreases so that the "admission period" of fuel is a function of the piston position. This initial idea of combustion by Rudolf Diesel resembles today's conventional diesel process.

He accurately describes the advantages of his thermodynamic "diesel" process in a practical application – only with his theoretical expectations. He summarizes the disadvantages of existing engines as follows: Energetic cooling of the cylinder or oven wall is essential to prevent overheating and keep the engine running. In addition, the expansion is insufficient to lower the exhaust gas (working gas) temperature. Consequently, the energy loss reduces the efficiency of the system. He also referred to the auto-ignition of petroleum in premixed conditions as a problem. Finally, he described the disadvantages of uncontrollable pressure and temperature rise of the direct injected spark ignition combustion process.

Further descriptions include the use of various fuels in the states: liquid, gaseous, and solid. He also outlined the combination of fuels to obtain a combustible mixture for his process. His book "Theorie und Konstruktion eines rationellen Wärmemotors zum Ersatz der Dampfmaschinen und der heute bekannten Verbrennungsmotoren", published in 1893, completed his work and provided further knowledge [3].

M. S. Wahl, *Emission Reduction with an Alternative Diesel Combustion Process*, Wissenschaftliche Reihe Fahrzeugtechnik Universität Stuttgart,
https://doi.org/10.1007/978-3-658-42094-9_1

In the same year, a second patent with the number DRP 82 168 was issued describing the thermodynamic diesel cycle as a combination of the Carnot [4] process and isobaric heat transfer for combustion. In the following years, he published further patents in France (243531 in December 1894 [5]), Germany (86633 in March 1895 [6]), the United States (608845 in August 1898 [5]), and additionally in Belgium, England, Switzerland, Luxemburg, as well as several other countries. His theoretical knowledge of thermodynamics and the will to build this "new" engine (first run in 1894 [7]) characterized the early beginnings of the diesel engine.

One hundred thirty years after Rudolf Diesel's patent, the diesel engine is still highly relevant in various sectors of the economy. However, in recent years, the automotive sector diversified, and the percentage approval of different types of engines changed. In Germany, a total of 15,225,000 passenger cars with diesel engines were registered in 2018. At that time, the price of a liter of diesel was 128.9-euro cents, while the last published average price on October 18, 2022, had reached a value of 195.13-euro cents, corresponding to an increase of 51 % in four years. For the first time since 1972, diesel fuel was more expensive than gasoline in Germany. Total registrations of diesel cars have declined slightly since 2018. As of January 1, 2022, approximately 14,824,000 diesel passenger cars were registered. In addition, there were 1,669,051 hybrid vehicles registered, in which gasoline and diesel engines are not separated. This results in a number of over fifteen million registered diesel passenger cars in Germany in January 2022, with an average age of 10.1 years. [8–10] Moreover, the average age of towing vehicles increased from 28.9 to 30 years between 2013 and 2022. A comparison of Roland Berger showed that diesel accounted for less than 3 % of new passenger car sales in the United States, Brazil, China, and Japan in 2014. However, the market for diesel passenger cars in India (52 % of newly registered passenger cars) and Europe (53 %) is different. Countries such as Italy and France had more than 53 % of newly registered passenger cars with diesel engines. [11]

Comparing the currently most efficient cars in each of the categories of diesel and gasoline vehicles regarding their fuel consumption during a real driving cycle and the approximately resulting CO_2 emissions, the following values are obtained:

- Diesel: Audi A2 3L (185) 3.6 l/100 km≈96 gCO_2/km
- Gasoline: Toyota Yaris Cross (99) 4.6 l/100 km≈109 gCO_2/km

The consumption is averaged over the number of vehicles indicated above in parentheses. The actual consumption for 185 Audi A2 3L is 3.6 liters of diesel per 100 km, corresponding to an emission of 96 gCO_2 per kilometer. The most fuel-efficient gasoline passenger consumed an average of 4.6 liters per 100 km across all 99 vehicles tested. This is equivalent to 109 gCO_2 per kilometer. [12] These results indicate the possibility of reducing CO_2 emissions for smaller vehicles, but the diesel engine also has advantages in larger, heavier vehicles.

The author would like to expressly point out the global application of combustion engines in existing vehicles as well as in the production of new vehicles. Vehicles produced today can be used for several decades on average. Therefore, a holistic assessment must replace the current calculation of CO_2 equivalent emissions based on conventional considerations. Furthermore, integrating new concepts for the propulsion of mobile and stationary energy converters today requires a highly precise analysis of the overall system and its performance (for example, PHEV). Related literature [13, 14].

In the European Union, the standard EN 590 describes the requirements and test methods of diesel fuel mixtures for motor vehicles. It is of great importance that diesel blends at service stations meet these requirements. Among other requirements, the standard specifies the minimum required cetane number (CN), which is set to 51. The maximum permissible content of polycyclic aromatic hydrocarbons and biodiesel (FAME – Fatty Acid Methyl Ester) is 8.0 % and 7.0 %, respectively. Furthermore, ingredients such as manganese, ash, sulfur, and water are regulated, and properties such as flash point, corrosive effect on copper, and oxidative resistance are legislated. [15]

In the United States, diesel is regulated by the Environmental Protection Agency (EPA), and different standards were implemented for the use as highway diesel (cars and trucks), non-road diesel (farm, construction, and mining), locomotive diesel, or marine diesel. In this regard, “Diesel engines and vehicles make up about a third of the entire transportation fleet in the U.S.”. Therefore, ultra-low sulfur diesel for highway use was implemented in 2006. [16] As a result, engine control units (ECUs) nowadays must be adjusted worldwide to comply with various regulations and legislations regarding fuel composition. However, not only the injection strategy nor the exhaust gas aftertreatment system is affected by global deviations of diesel fuel compositions. Also, the combustion process and, consequently, the engine development in terms of mechanical stress, combustion chamber geometry, or the injection system has to be adapted to meet the fuel properties.

The present work, in addition to two further projects at the IFS University of Stuttgart, addresses topics related to diesel injection, homogenization, and ignition delay. The focus of the first project, “Kraftstoffkennzahlen I” and the second project, “Kraftstoffkennzahlen II” was on homogeneous charge compression ignition (HCCI) and the correlation of physical and chemical ignition delay for different fuel compositions. The correlation of CN and low-temperature combustion was investigated in a BASF engine and single-cylinder Mercedes Benz engines OM646 Evo and OM642. Therefore, it was necessary to vary and adjust the piston geometry to optimize the combustion chamber geometry for HCCI. Furthermore, fuel preparation with an atomizer for intake manifold injection was optimized. Finally, a single-stage Arrhenius approach was adapted to predict the ignition delay of HCCI’s low and high-temperature combustion.

The new e-Skyactive D engine, which Mazda launched, will be available in 2023. A piston bowl geometry is adapted that enables partially premixed diesel combustion at low load and engine speed but also has a wide range of load and speed at conventional diesel combustion. The 3.3-liter in-line six-cylinder diesel engine uses distribution-controlled partially premixed compression ignition (DCPCI). Despite a lowered compression ratio of 14:1 and an egg-shaped dual-zone combustion chamber to separate the air-fuel mixture into two zones, a thermal efficiency of over 40 % can be achieved. This results in an extended engine map for the alternative DCPCI combustion process. Combined with a maximum rail pressure of 2500 bar and a low hybridization (153 Nm electric power) from a 48 Volt system called M Hybrid Boost. The overall fuel consumption in the WLTP cycle is 4.9 l/100km or 127 gCO_2/km in a Mazda CX60 e-Skyactive D. [17] Thus, the main difference to the conventional diesel combustion chamber is the piston geometry. This indicates that the premixed charge compression ignition (PCCI) process is valid to offer emission reduction opportunities for transient engine conditions. But therefore, the ignition delay and combustion must be better understood for this to be possible.

1.1 Conventional Diesel Combustion

Recently, different combustion chamber geometries for indirect and direct injection diesel engines have been used in four-stroke small and medium-sized engines. This development ranges from engines in which the combustion chamber is designed with a swirl or a turbulent pre-chamber (indirect injection – IDI) to engines with direct injection (DI) with a single combustion chamber. Here (DI), the piston bowl is shaped as an omega and forms the combustion chamber, as shown in **Figure 1.1**. The fuel injection design is performed to penetrate the fuel jets into the piston bowl at conventional diesel strategies (injection near TDC). Several concepts are conceivable, each generating a different jet breakup and atomization. The injection of highly pressurized diesel fuel into this combustion chamber with a direct injector at around TDC follows in high penetration depth.

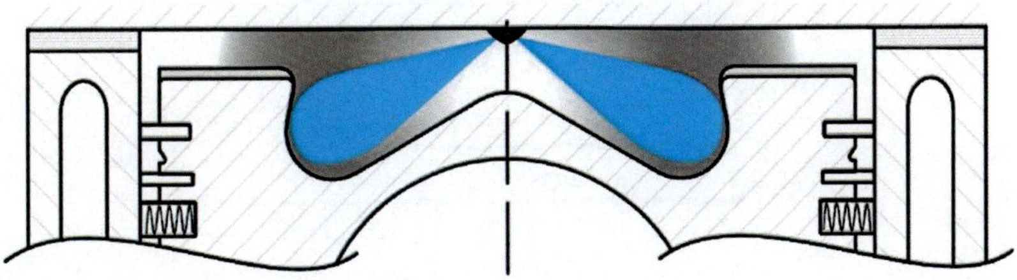

Figure 1.1: Conventional Diesel Injection. Piston Cross Section with Two Injection Coils in Blue [18].

The start of injection (SOI) is around the top dead center (TDC) for conventional diesel injection. The ignition delay (ID) is the time starting at SOI and ending with the start of combustion (SOC). The SOC can be detected as a change in the slope of the in-cylinder pressure or the heat-release analysis. [19] This heterogeneous mixing process forms a fuel and air composition that self-ignites at high temperatures and pressure during the compression stroke. For this process, combustion can be divided into two phases: premixed (or rapid mixing) and diffusive [19, 20]. While the physical and chemical ignition delay dominates the premixed phase, the diffusive phase is mainly characterized by the physical ID. The physical ID depends on jet propagation, atomization, and vaporization; therefore, the chemical ID correlates with fuel pre-heating and chemical pre-reactions. When the number of reactants is sufficient, the self-ignition process starts in the gas phase. For long-chain hydrocarbons, a single or two-stage ignition behavior is common and described in chapter 1.4.

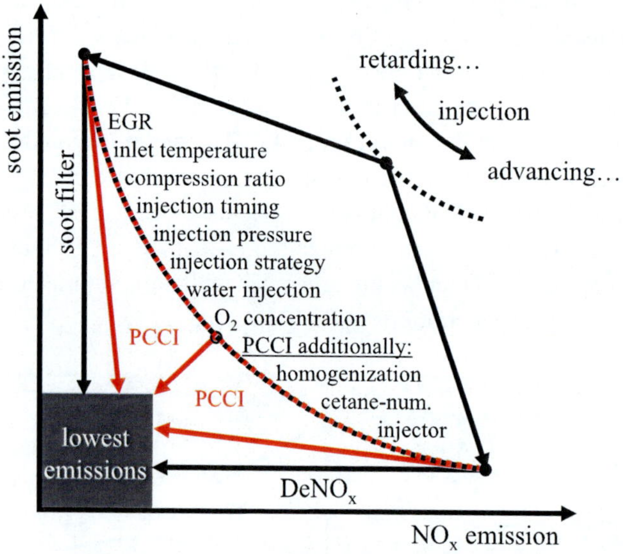

Figure 1.2: Schematic Soot and NOx Trade-Off in a Diesel Engine and the Benefits of PCCI (Adapted from [2] and [21]).

Figure 1.2 symbolizes the raw-emissions trade-off between soot and nitrogen oxides, primarily known in diesel engines. The raw emission of soot and NO_X hereby mainly depends on the injection timing. For advanced timing, the NO_X emissions increase but soot emissions decrease. Otherwise, retardation leads to lower NO_X and higher soot emissions. Therefore, soot and/or NO_X reduction systems are used in the exhaust gas aftertreatment. For soot, a particulate filter is usually utilized.

On the other hand, a NO_X storage catalyst (Lean NO_X Trap – LNT) or selective catalytic reduction (SCR) is applied for nitrogen gas. Since different exhaust gas aftertreatment systems have advantages and disadvantages, these must be properly evaluated. SCR, for example, has the disadvantage that a reducing additive (usually ammonia in an aqueous solution) must be injected. This requires additional peripheral equipment: a tank socket, a tank, and a dosing unit. On the other hand, the LNT, which is very close to the three-way catalyst (TWC), has the disadvantage of the duration for loading. This can take up to several minutes, and the combustion must then be operated sub-stoichiometrically for a few seconds. This regeneration of the catalyst thus involves energy expenditure.

Specifically, many parameters affect the formation mechanism of these emissions, where some can be seen in chapter 5. However, unlike the injection timing, most parameters have different effects on the formation mechanisms of soot and NO_X. As shown in **Figure 1.2** by the Pareto front in red (a set of most efficient solutions named after Vilfredo Pareto [22]), PCCI is predicted to reduce raw emissions using homogenization and low-temperature combustion [23]. Further advantages for the combustion process regarding efficiency and emissions (gasoline engine [24, 25], diesel engine [26]) can be achieved, for instance, with water injection, recirculated exhaust gas, better homogenization, fuel composition, and others.

Figure 1.3 illustrates a diesel injection jet as a sectional image for diffusive combustion. Here, a single injector jet is shown with the tip on the left side of the figure. The spray consists of black-colored liquid fuel and a rich air-fuel mixture represented by the white areas. The diesel fuel enters the combustion chamber on the left side at high pressure and temperature (for injection around TDC). The approximate timing of this schematic diesel fuel structure is highlighted in **Figure 1.4**, with the gray line at the main injection.

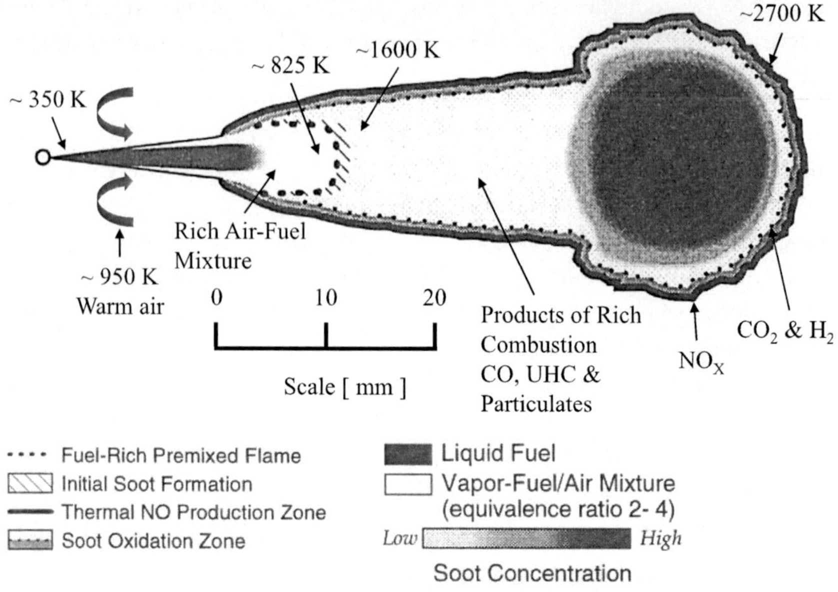

Figure 1.3: Schematic Structure of a Diesel Injection Jet from Laser-Sheet Imaging with Additional Information to Temperatures, Emission Formation Areas and Mixture (Adapted from [27] and [28]).

As **Figure 1.4** shows in a schematic form, the high variability in needle lift leads to possible adjustments and applications. In this figure, two pre-injections, one main injection, a close post-injection, and one late post-injection represent a potential injection pattern. Engine performance, noise, or raw emissions can be influenced over a wide range by varying the injection strategy. The early injections pre-condition the combustion chamber, the subsequent pre-injection directly affects the combustion noise, and the main injection is decisive for the net torque. On the other hand, early post-injections have the objective of combusting before exhaust opens, but not to perform work. This increases the exhaust gas temperature and, thus, the oxidation of soot. Late post-injection promotes unburned fuel into the exhaust gas. This serves, for example, as a reducing agent for the exhaust gas aftertreatment systems as described for LNT. Neither near post-injection nor late post-injection was used in this project. For conventional diesel measurements, one pre-injection was set. The PCCI injection strategy was performed with one up to three (very) early pre-injections and one main injection. [21]

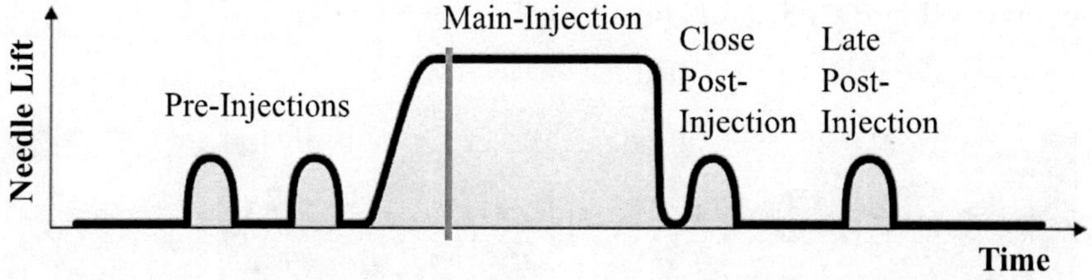

Figure 1.4: Needle Lift over Time: Schematic for a Multiple Diesel Injection Strategy with Pre and Post-Injections (Adapted from [29]). Gray Line to Indicate the Momentary of **Figure 1.3**.

1.2 Alternative Diesel Combustion

The last decades of engine development have been characterized by combining theoretical gasoline spark ignited (SI) and diesel combustion processes to reduce engine-out raw emissions and increase the engine performance for both engine types. As described in section 1.1, the feature of the conventional diesel process is the main injection around top dead center (TDC). This injection is followed by physical and chemical ignition delay, SOC (premixing), and diffusive combustion. This combustion process can be compared with charge stratification in gasoline engines, where local lean and rich areas are separated.

Most alternative diesel combustion processes focus on homogenization. They are summarized as controlled auto-ignition (CAI) combustion processes [30] with a pronounced low-temperature heat release (LTHR), which significantly reduces combustion temperatures. However, the timescale of this homogenization process exceeds the total injection time [20]. Therefore, it must be considered that the end of injection and SOC do not overlap. Moreover, an infinite time scale could theoretically lead to perfect homogenization. From these results, several alternative diesel combustion processes have been developed, which can be distinguished mainly by injection strategies (injection timing and injector position), homogenization, and the rate of recirculated exhaust gases (see **Figure 1.5** and **Figure 1.6**).

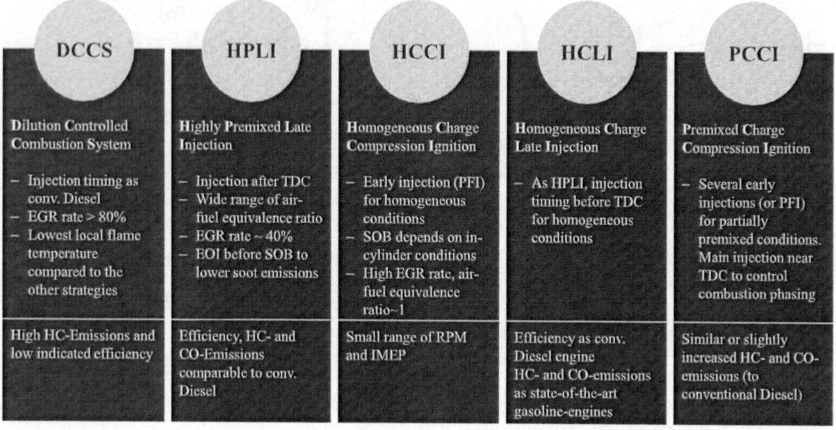

Figure 1.5: Comparison of Alternative Diesel Combustion Injection Strategies [1, 21, 31].

Many acronyms for alternative diesel combustion systems are described in the literature. For some, the methodology is similar in terms of exhaust gas recirculation (EGR), injection system, injection strategy, and other relevant parameters, but different acronyms are used. **Figure 1.5** briefly overviews several alternative combustion strategies selected based on the injection strategy. These five are discussed in more detail in the following sections.

In contrast to "hard", noisy conventional diesel combustion due to a large proportion of premixed combustion influenced by ignition delay and fuel evaporation, dilution by very early injection combined with a high EGR rate can reduce auto-ignition behavior and rapid pressure rise. Contrary to the trade-off between generating soot and NO_x in a conventional diesel engine, the high amount of EGR and, thus, low oxygen proportion does not enhance soot formation due to better homogenization by increased ignition delay [32]. In 1996, Takeda et al. published a paper with a strategy called "PREmixed lean DIesel Combustion" (PREDIC), using a medium-size engine with a single-cylinder displacement of 2004 cm^3.

They inserted a central injector and two side nozzles and used them individually and in parallel for the studies. Nitrogen oxides and soot reduction were "dramatic", but only minimal influences by swirl number and side nozzles were found [33]. This means that a single injector is sufficient to achieve significant improvements. Another paper by Kengo Kumano and Norimasa Iida [34] describes the influence of homogeneous or inhomogeneous mixtures in HCCI conditions. Different injection strategies were investigated to achieve a homogeneous or inhomogeneous mixture at SOC. The summary of the results leads to the following assumptions: The homogeneous condition results in a higher fluctuation of peak pressure but a lower variation in the time of occurrence (SOC) of LTHR and high-temperature heat release (HTHR). In summary, the influence of PREDIC on emission reduction is immense under different boundary conditions and mixture formation processes. Thus, the in-cylinder condition in the compression stroke controls the combustion behavior.

Figure 1.6 is an extension of **Figure 1.5** to show the injection timing of the different strategies. They all have in common that the EGR rate can be varied to control combustion phasing, so the ID is directly affected (see **Figure 4.4**). Combustion stability and controllability are most relevant for low emissions over all cycles. Therefore, the amount of recirculated gas should not exceed the incineration limits of the overall system.

In the diagram, two pressure curves are plotted against the crank angle. These two unfired operation curves are measured with different inlet pressures. In addition, a plot of the valve lift curve without valve clearance is in the upper right-hand corner. The negative valve overlap leads to intermediate compression in the gas exchange cycle. The valve opening position can be changed with an intake camshaft adjuster in the serial engine. For measurements in PCCI operation, however, high residual gas content is advantageous since the external exhaust gas recirculation is limited by the existing pressure conditions (p_2/p_3). Negative valve overlap is unsuitable in the higher load range since hot residual gas remains in the cylinder and does not participate in combustion.

For part-load operation, however, this results in a minor disadvantage, which is why all measurements were performed with the same camshaft setting.

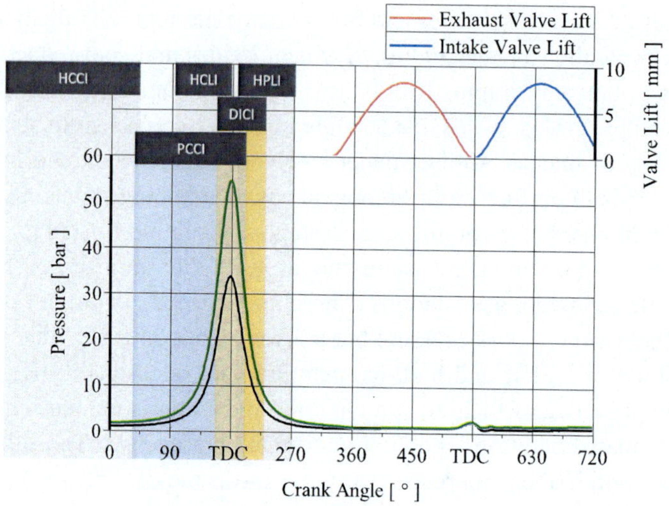

Figure 1.6: Alternative Diesel Combustion Strategies via the Injection Timing [1, 31] (cf. [35]).

To classify the process in its combustion mechanism, the triangle in **Figure 1.7** shows the most important forms of ignition and the associated progression.

The ignition of fuel always starts in local premixed areas in which, depending on the combustion process used, the energy input from outside (spark plug, combustion chamber temperature, pressure) leads to continuous flame propagation. The local distribution and time scale thereby determine the global burn-rate function. The fastest combustion of space ignition, where combustion starts at many points in the combustion chamber simultaneously, has the largest rate of heat release (ROHR). Damage prediction of engine components due to high maximum pressures and pressure gradients is not easy to handle, which makes space ignition difficult to control. In contrast, heat release in mixture-controlled combustion is uniform and controllable via the fuel mass flow thru the injector. The fuel mass flow introduced is almost proportional to the rate of heat release. A distinction between mixing-controlled and homogeneous compression ignition can be made in partially premixed diesel combustion. Although the premixed portion burns "partially homogeneously" by auto-ignition, the ROHR strongly dependends on the conditions in the cylinder. The main injection ideally burns "on the remaining fuel" in premixed combustion and then diffusively. Since the average mass temperature is locally too low to oxidize the soot formed at the end of combustion, diffusive combustion must be kept as low as possible (cf. chapters 1.3 and 3.1).

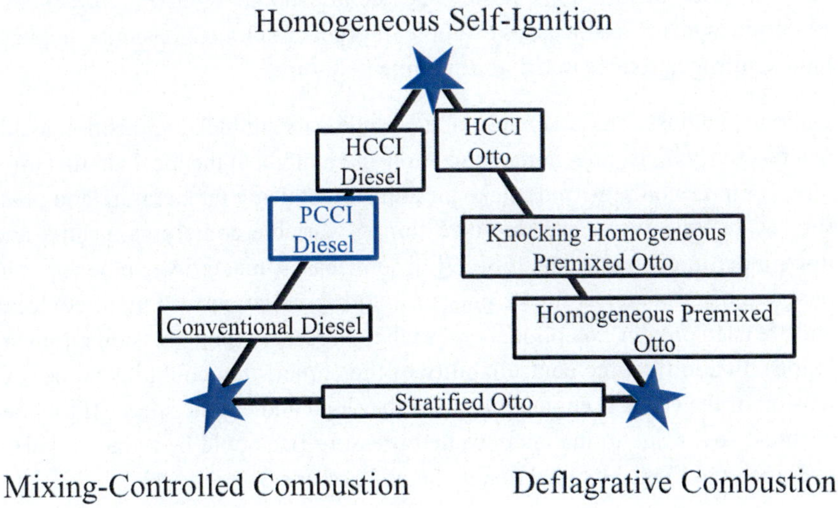

Figure 1.7: Classification of Gasoline and Diesel Engines with Regard to Flame Propagation and Combustion (Adapted from [36]).

1.2.1 HCCI

Homogeneous **c**harge **c**ompression **i**gnition uses the possibility of the self-ignition of a fuel with an oxidant (air) after compression (depending on the conditions in the cylinder). For diesel engines, the auto-ignition process is obvious. For the combustion process of a gasoline/spark ignition engine, HCCI has been developed to provide further advantages such as de-throttling (reduction of pumping losses), caloric advantages due to low-temperature combustion, and rapid heat release (an increase of thermodynamic efficiency) [37]. This is a disadvantage due to the controllability of the SOC and the comparatively small engine map area. HCCI can lead to comparably low NO_X and soot emissions and high efficiency in both diesel and gasoline engines. However, even without exhaust aftertreatment, no fuel consumption benefits are seen on the test bench. Otherwise, if the exhaust aftertreatment could be reduced by better engine-out emissions of the HCCI combustion, a reduction in fuel consumption is possible since, for example, no fuel enrichment has to be made to reduce nitrogen oxides in the aftertreatment system.

Options to adjust combustion phasing include varying fuel composition to adjust reactivity, EGR rate, homogenization and lambda of the local air-fuel mixture, compression ratio and intake air adjustment using temperature and pressure, water injection, variable valve timing, variable compression ratio, and in-cylinder injection timing. [38, 39] In some cases, reactivity can be continuously adjusted (such as intake pressure). Other variations are more complex and are used only in exceptional cases when necessary (compression ratio variation). In addition, the controllability of the combustion could lead to an extension of the (HCCI) engine map to higher load and engine speed. If the mechanical stress due to the uncontrolled pressure rise could be reduced, HCCI operation could be extended to a wide range of operating modes.

The intention of HCCI is to achieve near-optimized homogenization of fuel, air, and recirculated exhaust gases throughout the combustion chamber at SOI. The result is a two-stage heat release with fast reactivity. However, the flame front known from conventional diesel combustion becomes global ignition. [38] Lower particulate matter (PM, soot) and NO_x emissions compared to conventional diesel are beneficial in HCCI combustion due to local lean areas and overall lower combustion temperature (Zeldovich mechanism). Emissions of carbon monoxide (CO) and total hydrocarbon (THC – "total mass of open chain and cyclic hydrocarbon molecules" [40] including CH_4 – measurable by a flame ionization detector, FID) are high but can be treated by catalysts. [41]

1.2.2 HPLI

Highly **p**remixed **l**ate **i**njection is based on conventional diesel combustion geometry (mainly piston). The engine map combines HPLI (low load and engine speed) with conventional diesel combustion (higher loads and engine speed). Furthermore, the spray geometry of the direct injection engine is combined with the omega bowl in the piston. SOI is shifted after TDC, and the EGR rate is used to retard the SOC. The main objective is to retard the first heat release after the end of injection (EOI) for charge homogenization. High EGR rates and homogenization at SOC results in low NO_x and particulate matter emissions. [31]

1.2.3 HCLI

Homogeneous **c**harge **l**ate **i**njection uses, such as HPLI, almost the same engine parts known from conventional diesel. This allows a wide engine speed and load range with different combustion strategies. For HCLI, the injection occurs around 40 °CA before TDC (°CAbTDC), resulting in comparably high premix ranges in the combustion chamber before the first heat release. This injection strategy is similar to HCCI but has a shorter ID time and, thus, less homogenization. [31]

1.2.4 DCCS

A **d**ilution-**c**ontrolled **c**ombustion **s**ystem is the most excessive alternative to conventional diesel combustion. A high EGR rate (about 80 %) results in low local and global in-cylinder temperatures below the formation temperature of NO_X. Hence, diffusive combustion is applicable because chemical reactions to form soot are prevented. The flame lift-off increases (more time for premixing and lowering soot), and the local average equivalence ratio at SOC remains almost constant. The disadvantage is the formation of hydrocarbon and carbon oxide emissions, as these are not oxygenated in the bulk gas. [42–44]

1.2.5 PCCI

A way to combine the advantages of the abovementioned combustion strategies is **p**artially **p**remixed **c**ompression **i**gnition with early pre-injection for homogenization and one main injection. Therefore, counteracting HCCI, HPLI, HCLI, and DCCS, the main injection allows combustion phasing. The SOI of the main injection varies with EGR rate, engine speed, injection pressure, amount of premixed combustion, and ongoing low or high-temperature heat release (bulk gas temperature). Combustion phasing through a slightly reduced compression ratio and high EGR rates results in ultra-low soot and NO_x emissions. However, controlling combustion phasing is still possible in a wide operating range.

1.3 The Disadvantage of Conventional Diesel Combustion on Soot and Nitrogen Oxide

The first step is to classify soot and particles properly. While soot is merely a carbon compound, particles can be fuel or oil deposits, ash, metal, corrosion, or sulfate. Therefore, it is at least relevant to have a superficial understanding of the complex soot formation to identify the benefits of alternative diesel combustion processes and, thus, the chemical process of carbon in a lean air-fuel environment.

A different approach is shown by Kamimoto et al.; their publication describes the possibility of high-temperature combustion for the reduction of nitrogen oxides and soot. In this process, the local formation of nitrogen oxides is just suppressed, and the resulting soot is oxidized at temperatures above 1900 K.

In the diagram, this high-temperature combustion can be placed in the area for conventional combustion but is not in the range for soot or nitrogen oxide formation. This very small operating range limits the application significantly for series use and was not pursued further. [45]

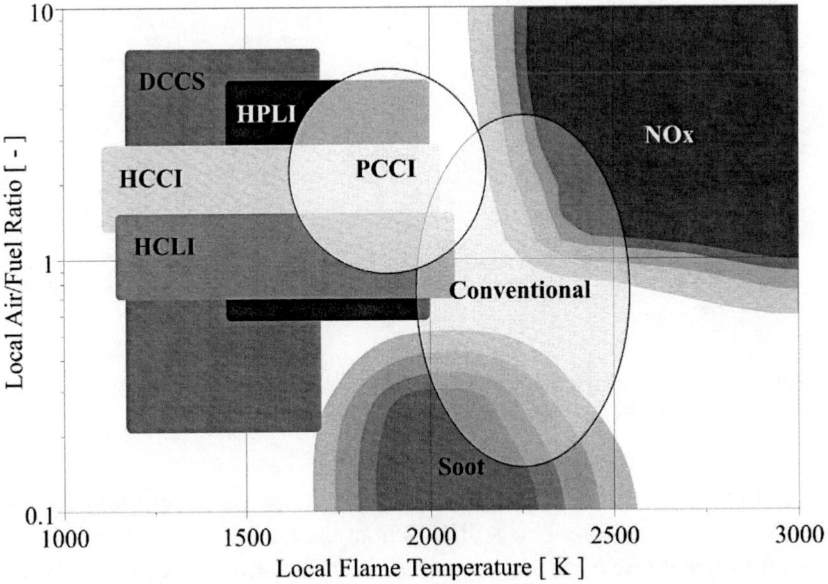

Figure 1.8: Classification of Conventional and Alternative Diesel Combustion Processes on a Local Soot and Nitrogen Oxide Formation Chart Depending on Local Lambda and Local Flame Temperature (Adapted from [31, 46]).

<u>Soot:</u> The formation of carbon particles has not yet been fully clarified. Nevertheless, a subdivision into organic and inorganic fractions is reasonable. PM is understood as the sum of all particles – i.e., the organic and non-organic fractions. However, the proportion of organic compounds and elemental carbon is responsible for most particle mass.

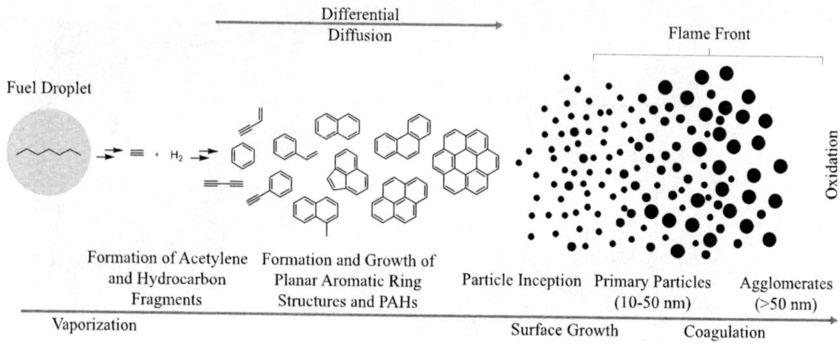

Figure 1.9: Simplified Formation Mechanism of Soot over Time (Adapted from [47–49]).

The formation of soot begins in locally rich areas, as shown in **Figure 1.8**. In most cases, the fuel droplet is still present at the beginning of combustion. A combustible mixture of air and vaporized fuel then forms around the fuel droplet, as shown in **Figure 1.3**. Once this burns in a premixed combustion and the area surrounding the fuel droplet is heated, the chemical processes for the formation of soot begin. Although the formation of new particles (left part in **Figure 1.9**) is responsible for only a small proportion of the total soot, the primary particles for subsequent surface growth are formed from them. Failure to form acetylene and hydrocarbon fractures would directly result in less soot formation. The fuel droplet partially changes its structure at the molecular level and forms PAH's. This subsequently results in the formation of larger aromatic rings, which then form soot particles. Chapter 3.1 describes the measuring principle and the necessary adjustments to the engine test bench. Unlike soot formation in locally rich areas, oxidation with oxygen can reduce particulate levels (cf. [50] and [51, 52]):

- Low oxygen content limits soot oxidation. However, CO oxidation is impaired prior to soot oxidation.
- When the oxygen content is sufficient, the oxygen concentration no longer affects the intensity of soot oxidation.

- Start of soot oxidation with a time delay (induction time of 2–3 ms) after the temperature of 1200–1300 K is exceeded.
- The lower temperature limit is at approx. 1200–1300 K. For higher conversion rates (>60 %) temperature above 1500 K is required. Above 1700 K, almost complete oxidation is possible.

Oxides of Nitrogen: As **Figure 1.8** introduced, the conventional diesel injection strategy faces the difficult conflict of forming soot or nitrogen oxides. The ability to operate between the formation mechanisms of soot and NO_X with diffusive combustion is not controllable. For an injection advance, the local temperature at air-fuel lean operation conditions results in high local combustion temperature and therefore NO_X formation. In section 3.2, the formation mechanism and the measurement at the test bench are described.

Reasons for using EGR for NO_X abatement [53, 54]:

- Overall reduction of exhaust gas mass flow.
- The reduction of the local air-fuel ratio λ.
- The lowering of the maximum temperature by reducing the flame speed.
- Additionally: Less gas movement with an open swirl flap reduces the mixing speed and, thus, the flame speed. This also leads to better PCCI performance by increasing ID.

Nitrogen oxides cannot be reduced via a three-way catalyst due to the excess oxygen in the quality-controlled diesel process. A NO_X storage catalyst or selective catalytic reaction with urea is suitable for reducing nitrogen oxides in the exhaust gas of high-speed diesel engines. Depending on the application, lean combustion results in an exhaust gas temperature below the limit for diesel engine exhaust gas aftertreatment. Thermolysis of the urea to ammonia in the SCR system requires approx. 170 °C, while storage of the nitrogen oxides in the storage catalyst requires 250 °C. In addition, when using an LNT, the combustion process must repeatedly be sub-stoichiometrically controlled to effect the reduction of NO_X in the storage catalyst to elemental nitrogen. [55]

1.4 Specification of the Two-Stage Ignition Delay

Under homogeneous conditions in the combustion chamber, diesel fuel leads to a pronounced two-stage ignition. In literature, these two stages are referred to as cool flame and hot flame. The associated onset of combustion is divided into LTHR and HTHR. In between, slowed chemical reactions result in a negative temperature coefficient.

Dietrich points out that mixture formation must be completed as far as possible (local lambda above 0.6 and mixture temperature below 1400 K) before the combustion temperature rises further. This leads to the absence of heavy soot formation by bypassing the soot formation mechanisms via low-temperature combustion. [56]

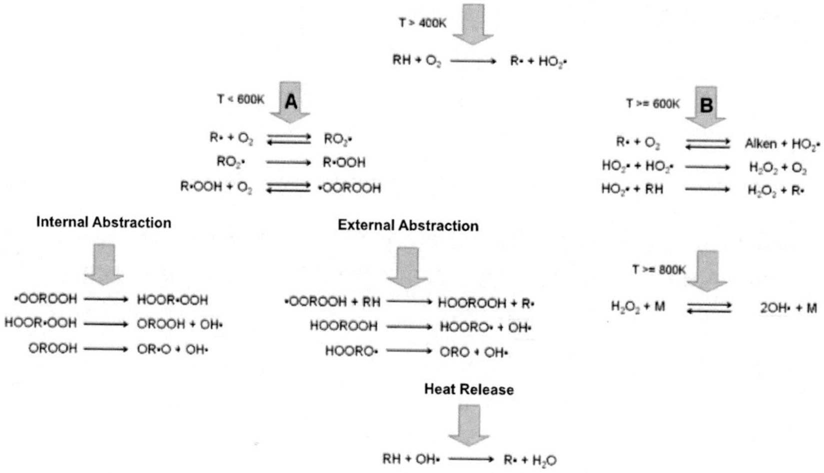

Figure 1.10: Low Temperature Reaction Pathways at Start of Combustion (Adapted from [57]).

A variety of reaction pathways characterizes the combustion of diesel. The composition of the fuel and the locally occurring composition determine the combustion process. Diesel is usually replaced by n-heptane for the studies describing the two-stage heat release. N-heptane belongs to the group of alkanes and exhibits a single-stage ignition above 900 K. Below that, two-stage combustion forms in the pressure-temperature diagram. This low-temperature range is often referred to as cool flame. Rether has summarized the temperature-dependent reaction paths of the combustion of n-heptane from various sources, as adapted in **Figure 1.10** [57]. In conventional diesel combustion, the high temperature prevails even at the start of injection due to compression in the combustion chamber.

Therefore, the ignition delay is mainly determined by physical processes. Injection leads to jet dispersion, jet breakup, and the disintegration of the droplets. In conventional injection, the chemical ignition delay – superimposed in time – leads to a first premixed combustion. At this point, the injection valve is usually still open, and diffusive combustion continuously occurs. The rising temperature in the end gas and the rising mass mean temperature follows in a single heat release. On the contrary, an early injection for premixing is less influenced by physical ignition delay. The conditions in the combustion chamber mainly determine the chemical ignition delay and subsequent combustion. As shown in **Figure 1.10**, the first radical formation occurs at about 400 K onwards. Further reaction is possible via path A or B, depending on the temperature. After internal or external abstraction, initial heat release is possible via reaction path A. A fuel radical and water is formed. Up to about 800 K, however, this reaction path aborts, and no further exothermic reaction occurs. The actual high-temperature reaction begins at about 1100 K. The previously formed carbon monoxide strongly oxidizes exothermically to carbon dioxide. This effect, in conjunction with the low turbulence due to lack of swirl and quenching at low speed, partly also leads to incomplete oxidation of CO. If this reaction fails to occur, the combustion efficiency drops considerably.

The two stages of self-ignition were already described by Jost [58] in 1939 and Sitkei [59] in 1964. **Figure 1.11** is intended to describe the dependence of auto-ignition on behavior of temperature in the unburned mass and the system's pressure. These investigations were not carried out with diesel fuel but with a diesel component (the majority of published studies on ignition delay dependence refer to n-heptane). The data required for this purpose can be obtained in a shock wave tube. However, compression apparatuses are too slow to map chemical ignition distortions, and possible pre-reactions lead to incorrect measurement results. For the measured data in the figure, there is a complete absence of auto-ignition at a lower temperature, even under high pressure. For increasing temperature, the first stage of auto-ignition follows in the cold flame region (except for low pressure). This occurs due to a degenerative chain linkage, in which first long-chain hydrocarbons are oxidized to stable intermediates. Their decomposition starts gradually, but the resulting radicals and heat release accelerate this process disproportionately (right picture, middle section). This results in stable formaldehyde (blue flame in activated form) and free radicals, which promote the oxidation of HC. The oxidation of evolved carbon monoxide to carbon dioxide characterizes τ_3, which subsequently thermally explodes and triggers a strong pressure increase.

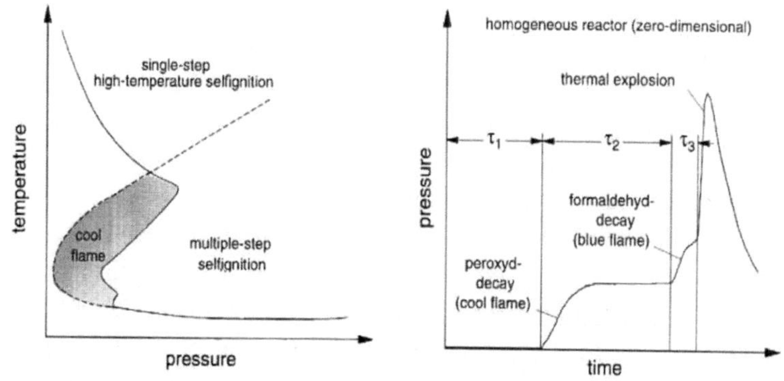

Figure 1.11: Reaction Kinetic Areas Depending on Temperature, Pressure and Time (As in [60] Adapted from [58, 59]).

The two-stage heat release: As the temperature increases in the low-temperature range, there is a continuous possibility that the peroxide concentration drops below a critical threshold. This reduces the radicals, and subsequent chemical reactions are thereby significantly decelerated. This area is characterized by decreasing exothermic reactions, which is why the dissipated heat shows a negative gradient. This is called the negative temperature coefficient (NTC) and depends on the engine speed due to time, temperature, and pressure-based reactions.

1.5 Empirical Ignition Delay Model

The empirical model, derived from the ancient Greek *empeiría*, is based on experiential findings. On the other hand, the phenomenological model describes the relationship of empirical relationships even more deeply. From this, derivations are formed, which are not derived from the theory but are based on the fundamental theory. The empirical model is thus usually formulated more simply but considers the entire system rather than the individual case. Applying this to the diesel engine's ignition delay model: the empirical model represents the simplest equation to predict the ignition delay approximately precisely for all eventualities. On the other hand, the phenomenological model considers the injection pattern, for example, to calculate evaporation more accurately. This is then incorporated into a higher-level model, which is based on empirical. The empirical model for determining the ignition delay is thus based on the findings from measurements and the general consideration of the correlations.

According to Barba [61], an injector was analyzed, which exhibits a delay of 0.3 ms between the start of electrical actuation and the start of injection, independent of the engine operating point. Thus, this is a constant factor in time, which, however, has a percentage effect on the total ignition delay. For short ignition delays, the time lag is therefore weighted proportionally higher than for long ignition delays. However, this empirical model takes account of this delay from the start of actuation to the start of injection, which is why the injector does not have to be analyzed separately for this factor. Thus, the model is independent of the injector type and must be adjusted exclusively to the engine measurements.

A comprehensive overview should be given to summarize the already published empirical ID models (cf. [62] adapted from [63]). While understanding interrelationships is essential in the early development phase, models are developed to describe these complex processes. These processes can be of various types, but understanding knocking combustion cycles (gasoline engine), combustion phasing, or ID are often in primary focus in such examinations. The tendencies, variabilities, and relations must be understood in detail to predict the knock dependency, combustion phasing, or ID. However, rapid prediction is required for the development process, and models are necessary for function development. To explain in the case of combustion phasing, boundary conditions such as the local and global air-fuel ratio, the air density, and the in-cylinder temperature affect the start of combustion, local flame temperature, knock-tendency, pressure gradients, and other combustion-related parameters. Empirical models are developed to predict combustion phasing by combining knowledge of the individual characteristics and simply describing the relationships in a single term. Obviously, this simplification leads to a lack of accuracy, but a well-developed model can be used as a tool for further investigations with predefined parameters.

As with combustion phasing, models are commonly used to describe the ID. However, in this case, the focus is on a self-ignition process – also called auto-ignition [64] – with a diesel blend. Since diesel has a two-stage ignition (cf. section 1.4), the start of low and high-temperature heat release must be predicted. Furthermore, the injection strategy has a relevant influence on the mixture and, thus, on the local conditions. This correlates with the premix quality and the tendency to self-ignition. In this context, the ID in a global and local lean premix area differs significantly from a global lean but locally rich combustion chamber (irregularities in the mixture).

To describe the temperature dependency of a systems ignitability, the so-called Arrhenius equation was first developed to predict reaction rates and was published in 1889 by Svante Arrhenius. Arrhenius thus started from van't Hoffs equation (published in 1884) and adjusted his equation to that given in Eq. 1.1. In his case, the system is not defined but must have thermally-induced processes or reactions. The Arrhenius equation:

$$k = A \cdot e^{\frac{-E_a}{RT}} \quad \text{Eq. 1.1}$$

consists of a pre-exponential factor A and an exponential term including the activation energy E_a, the gas constant R, and the (actual) process temperature T of the system. Following this, k is the rate constant or the frequency of collisions in a system. The activation energy is the minimum energy required by a chemical system to initiate a reaction. This applied activation energy/enthalpy is mandatory to start the combustion of the auto-igniting diesel. The required activation energy depends on different boundaries, such as local conditions or the fuel used. [65–69]

A modified Arrhenius equation takes the temperature dependency of the (frequency) factor *A* into account:

$$k = A \cdot T^n \cdot e^{\frac{-E_a}{RT}} \quad \text{Eq. 1.2}$$

For example, Wolfer, Bauer, and Elliot developed zero-dimensional ID models for ICEs based on the Arrhenius equation. Wolfer started with experiments in an incinerator and published his results in 1938. His results can be summarized in two dependencies: temperature and pressure. If one is increased, the ID shortens. [70] Bauer's research, published in 1939, was also executed at Cambridge University but on a single-cylinder test bench. Moreover, his results cover the Arrhenius expression, and also in his examinations, the pressure term is pre-exponential. In his work, the ID is referred to as ignition lag and is affected by the "compression ratio, intake pressure, cooling water temperature, load, injection advance, engine speed, and combustion-chamber design". [71]. Elliott divided the ID or ignition lag in advance into physical and chemical components. His equation has a more complex form with two exponential terms. [72] Subsequently, a large number of Arrhenius-based equations have been developed. While the main focus is on the extension to consider a more complex set of parameters. This refers to parameters such as the recirculation rate, the air-fuel ratio, or the injection pressure.

In a further step, the Arrhenius equation was adapted to the existing two-stage heat release of diesel fuel, where a single ID calculation is insufficient. Therefore, Weisser developed a three-stage Arrhenius approach in 2001, where he considered the negative temperature coefficient in his ID model [73]. While:

$$\frac{1}{\tau_{ID}} = \frac{1}{\tau_{ID,1} + \tau_{ID,2}} + \frac{1}{\tau_{ID,3}} \quad \text{Eq. 1.3}$$

represents the two stages of combustion and the negative temperature coefficient. Furthermore, an ignition integral of Eq. 1.4 with Eq. 1.3 leads to the ID of the low and high-temperature combustion.

$$\int_{t_{SOI}}^{t_{SOC}} \frac{1}{\tau_{ID}} dt = 1 \qquad \text{Eq. 1.4}$$

The calculation of the ignition integral begins with the start of injection and therefore depends on the injection system. The integration of the reciprocal τ_{ID} ends with the equal to one condition. This is equivalent to SOC and stops the integration immediately.

In literature, different ID models are available and adapted to various boundary conditions. However, the engine tests in this project were performed with a different number of injections as well as a main injection for pressure forming. For this purpose, the existing equations had to be further developed to consider the injection timing, the fuel quantity, and other conditions. The empirically based ignition delay prediction model of a PCCI injection strategy is presented in chapter 5.

2 Thermodynamic Basics to Evaluate PCCI Measurements

First of all, thermodynamic equations for a (diesel) engine will be discussed. Thereafter, the losses in PCCI-like combustion with respect to LTHR, NTC, and HTHR of the pre-injected fuel are in focus.

In this project, an investigation of partially premixed combustion with regard to the applicability and a way to develop an empirical ID model was sought. The engine test bench measurements are relevant as an experimental database to set up the model and to compare PCCI with a conventional diesel strategy. The input of this model is the SOI of the first injection, the SOC (LTHR and HTHR), and the main parameters that directly affect the combustion. Therefore, the in-cylinder temperature, pressure (measured), EGR rate (externally measured and internally calculated), and the global air-fuel ratio after pre-injections must be available for calculation. The difference in burn rate and ROHR shall be given in the first step. In the second step, the advantages and disadvantages of the burn rate and ROHR will be compared and evaluated to assess the PCCI combustion process.

Following the ideal gas law, the thermal equation of state is:

$$p \cdot v = R \cdot T \qquad \text{Eq. 2.1}$$

Multiplied with the systems mass, it follows:

$$p \cdot V = m \cdot R \cdot T \qquad \text{Eq. 2.2}$$

And in differential form:

$$p_c \cdot \frac{dV}{d\varphi} + V \cdot \frac{dp_c}{d\varphi} = m_c \cdot R \cdot \frac{dT_c}{d\varphi} + m_c \cdot T_c \cdot \frac{dR}{d\varphi} + R \cdot T_c \cdot \frac{dm_c}{d\varphi} \qquad \text{Eq. 2.3}$$

M. S. Wahl, *Emission Reduction with an Alternative Diesel Combustion Process*, Wissenschaftliche Reihe Fahrzeugtechnik Universität Stuttgart, https://doi.org/10.1007/978-3-658-42094-9_2

With R ($R_{Specific}$) as:

$$R_{specific} = c_p - c_v \qquad \text{Eq. 2.4}$$

further:

$$\kappa = \frac{c_p}{c_v} \qquad \text{Eq. 2.5}$$

The first law of thermodynamics and the thermal Eq. 2.3 are most important for calculating the combustion process. The first law of thermodynamics deals with the principle of energy conservation (and thus enthalpy), while the second law describes the irreversibility (change of entropy) in a counterclockwise or clockwise cycle. While the clockwise cycle converts heat into mechanical energy, the counterclockwise cycle raises thermal energy from a lower to a higher temperature level (refrigerator or heat pumps) [4].

Hence the enthalpy exchange is not relevant in the low-pressure cycle; the first law of thermodynamics is as follows:

$$\frac{dU}{d\varphi} = \frac{dQ_b}{d\varphi} + \frac{dQ_w}{d\varphi} + \frac{dH_l}{d\varphi} + \frac{dW}{d\varphi} \qquad \text{Eq. 2.6}$$

With the internal energy described with φ as the measured crank angle: Q_b as the released heat, Q_w the wall heat transfer, the leakage H_l and the work W. This work, or, in the reciprocal engine the volume work (change), can be written as:

$$\frac{dW}{d\varphi} = -p_c \cdot \frac{dV}{d\varphi} \qquad \text{Eq. 2.7}$$

To calculate the burn rate, rearranging Eq. 2.6 with Eq. 2.7 to the burn rate $\frac{dQ_b}{d\varphi}$ is necessary. Another definition of the burn rate is:

$$\frac{dQ_b}{d\varphi} = H_u \cdot \eta_e \cdot \frac{dm_f}{d\varphi} \qquad \text{Eq. 2.8}$$

This equation uses the lower heating value of the fuel, the fuel mass and the conversion/combustion efficiency. SGS Germany GmbH tested the red diesel used for this project under report number SP2-00688.001. The result of the report is a lower heating value of 42.918 MJ/kg, density at 15 °C 828.4 kg/m^3, nitrogen 14 ppm, sulfur 8.4 mg/kg, carbon 86.1 %, hydrogen 13.9 %, and oxygen at 0.00 %.

Furthermore, the measured in-cylinder pressure p_c is required for Eq. 2.7. Since the measurement principle for the in-cylinder pressure does not correspond to a relative pressure, the measured pressure needs a zero correction, as described in the section (see section 3.5).

For the burn rate calculation, assumptions are necessary to adjust the model parameters. In contradiction, the calculation of the ROHR is based on in-cylinder pressure, combustion chamber volume as in Eq. 2.13, heat capacity, and crank angle. The rate of heat release as a differential is:

$$\frac{dQ_h}{d\varphi} = \frac{\kappa}{\kappa - 1} \cdot p_c \cdot \frac{dV}{d\varphi} + \frac{1}{\kappa - 1} \cdot V \cdot \frac{dp_c}{d\varphi} \qquad \text{Eq. 2.9}$$

With $\frac{\kappa}{\kappa-1}$ from Eq. 2.4 and Eq. 2.5 and:

$$\mathrm{T} = T_0 \cdot \frac{p \cdot \mathrm{V}}{p_0 \cdot V_0} \text{ mit } T_0 = 360\,K \qquad \text{Eq. 2.10}$$

The change of mass in the cylinder for the diesel cycle is:

$$\frac{dm_c}{d\varphi} = \frac{dm_e}{d\varphi} + \frac{dm_i}{d\varphi} + \frac{dm_l}{d\varphi} + \frac{dm_f}{d\varphi} \qquad \text{Eq. 2.11}$$

(+/-) (-/+) (+/-) (-/+) (+)

where the exhaust, intake, leakage, and fuel mass flow are considered. The direction of flow determines the sign, whereby all incoming energies are positive, and all outgoing energies are negative. As for leakage area, the piston rings are in main focus, but the intake or exhaust valves could also be considered for a 4-stroke engine including intake and exhaust valves.

A relevant dimension for all in-cylinder calculations is the compression ratio (epsilon) defined in Eq. 2.12. It is therefore assumed that there is no mechanical slackness:

$$\varepsilon = \frac{V_d + V_c}{V_c} \qquad \text{Eq. 2.12}$$

The relevant order of magnitude for epsilon for gasoline engines is seven to eleven; for diesel engines, 14 to 24 is common. A difference of around one between the mechanical and thermodynamic epsilon exists due to leakage, wall heat transfer, geometrical deviations, and other factors. The formula can calculate the mechanical compression ratio if the combustion chamber geometry and the crankshaft radius are known. The thermodynamic epsilon is calculated with a measured unfired pressure curve. For this purpose, preconditioning the combustion chamber to compare the conditions in the cylinder notionally between unfired and fired operation is of great importance. The thermal expansion of the mechanical parts necessarily reduces leakage, gap volume, and wall heat transfer. For a given crank angle, the associated combustion chamber volume $V(\varphi)$ without cylinder offset is calculated as follows:

$$V(\varphi) = V_d \cdot \left(\frac{1}{\varepsilon - 1} + \frac{1 - \cos(\varphi)}{2} + \frac{1}{2\lambda_l} \cdot \left(1 - \sqrt{1 - \lambda_l^2 \cdot \sin^2(\varphi)} \right) \right) \qquad \text{Eq. 2.13}$$

With the calculated epsilon, the temperature can be estimated at the end of the compression via the following equation:

$$T_c = T_0 \cdot \varepsilon^{n-1} \qquad \text{Eq. 2.14}$$

This temperature leads to around 400 – 500 °C for gasoline and 700 – 900 °C for diesel engines, respectively. [74]

The indicated mean pressure is essential for determining combustion and actual performance. For this purpose, the indicated work per cycle is necessary:

$$W_i = \oint p dV \qquad \text{Eq. 2.15}$$

The definition for indicated work:

$$W_i = p_{m,i} \cdot V_d \qquad \text{Eq. 2.16}$$

And the indicated mean effective pressure (IMEP or $p_{m,i}$) is:

$$p_{m,i} = \frac{1}{V_d} \oint p dV \qquad \text{Eq. 2.17}$$

As can be seen from Eq. 2.17, the indicated mean pressure refers to the displacement volume. Therefore, engines with different displacements, compression ratios, or combustion chamber geometry can be compared directly with IMEP. A distinction must still be made between the overall IMEP and the high-pressure or low-pressure sections. The low-pressure (gas exchange) cycle is usually negative because work has to be introduced. However, the gas exchange cycle can also perform positive work if a mechanical compressor or the exhaust enthalpy is used with a turbocharger. The high-pressure cycle is positive for fired operation once the losses of the real engine are overcome. The investigated processes presented here range from IMEP 4 to 10 bar.

Summarized:

- Indirect indication values such as the burn rate, IMEP, coefficient of variation (IMEP), ignition delay, SOC, and energy conversion points (2 or 5 %MFB, 50 %MFB, 95 or 97 %MFB) are based on thermodynamic precalculations and correspond with the accuracy of the assumptions.

- Direct indication values such as the maximum in-cylinder pressure, phasing of the maximum in-cylinder pressure, maximum pressure rise, and derivative of the maximum pressure rise can be determined directly from the in-cylinder indication.

- By combining indirect and direct indication values, precise statements can be made regarding combustion.

These indication values are used in section 4.5 to compare conventional and PCCI combustion.

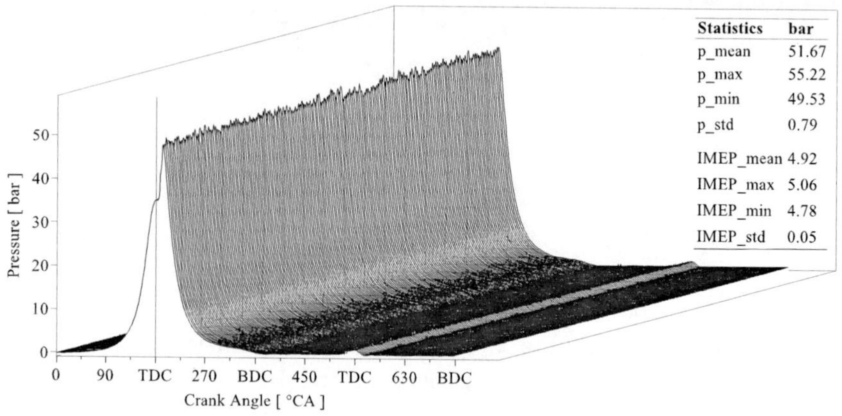

Figure 2.1: 199 Measured Cycles at 1500 min^{-1} and IMEP of 5.06 bar Plotted as Single Cycles.

Figure 12 shows 199 cycles to illustrate combustion stability. The standard deviation of the maximum cylinder pressure is 0.79, and that of the indexed mean pressure is 0.05.

It is usual to evaluate the combustion using various characteristic values. Among other aspects, these permit the quality of the combustion, the combustion stability, the noise of combustion (acoustic evaluation), and the position of combustion phasing to be determined. The following table gives an overview of characteristic values. In chapter 4.5, these are partially used to evaluate partially premixed and conventional diesel combustion. **Table 2.1** shows that various direct and indirect indication parameters are relevant for the assessment. Besides defined characteristic values such as the conversion rate or the burn rate, some are directly determinable and, therefore, applicable for a fast analysis on the test bench for the evaluation of combustion.

Table 2.1: Characteristic Values to Quantify Combustion (Compare [61]).

Characteristic Value	**Evaluation Criterion**
IMEP	Work Internal Efficiency Friction Losses
σ_{IMEP} – Standard Deviation of IMEP	Combustion Stability Smooth Running Misfire
p_{max} – Peak Pressure	Noise Mechanical Stress
$\alpha_{p,max}$ – Position Peak Pressure	Combustion Phasing / Efficiency
$\left(\frac{dp}{d\alpha}\right)_{max}$ – Maximum Pressure Rise	Conversion Rate Noise
$\left(\frac{dp}{d\alpha^2}\right)_{max}$ – Maximum Pressure Rise Rate	Noise

2.1 The Diesel Cycle and the Benefits of Thermal Efficiency

For a direct injection engine with omitted crevice flow (cf. [19]), the mass flow over the system boundaries after the intake valve closes consists only of injected fuel, as shown by Eq. 2.11:

$$\frac{dQ}{d\varphi} - p\frac{dV}{d\varphi} + \dot{m}_f h_f = \frac{dU}{d\varphi} \quad \text{Eq. 2.18}$$

The efficiency of the cycle is counteracted by increasing the recirculation of unburned and burned exhaust gas. However, unlike ideal stoichiometric combustion – assuming no fuel and no oxygen in the exhaust gas – the globally lean air-fuel ratio in a diesel combustion process has disadvantages in recirculation, as will be described in the next section.

On the one hand, the deposition of all types of particles in the exhaust pipe clogs the EGR cooler/heat exchanger, the valve, the intake pipe, and the intake port over time. The deposition over the engine's life is more constant than the (short-term) variations from cycle-to-cycle. This deposition results in inconsistent EGR valve control over the engine's life and even the individual cycle. The change in flow resistance (back pressure) affects the recirculated gas flow. Reduction of this deposition by oxygenation at high exhaust gas temperature is possible with different combustion modes.

On the other hand, the amount of gas required to achieve a high recirculation of inert fraction increases with increasing lambda in the cylinder. This is because the oxygen content (of an over-stoichiometric lambda) in the recirculation flow supports combustion instead of acting as an inert gas. In addition, as the EGR rate increases, the efficiency in the cylinder decreases due to the increasing specific heat capacity of the working gas, resulting in a lower pressure level. For example, at an EGR rate of 40 % instead of 0 %, there is an efficiency reduction of 0.6 % [75]. This disadvantage must be compensated by an increase in combustion efficiency to equalize the cycle efficiency.

For the combustion process in an open thermodynamic cycle, the loss of efficiency is unavoidable and irreversible at any time. The first part of the compression stroke has the lowest loss in a pre-heated engine. This loss between a thermodynamic ideal/perfect engine/cycle and the real engine contains [76-78]:

- Consideration of a real charge at IVC.
- Combustion phasing.
- Incomplete and imperfect combustion.
- Real combustion: The appearing heat release in shape and duration.
- Real gas: calorific properties as a function of pressure, temperature, and composition.
- Wall heat transfer.
- Incomplete expansion (EVO).

- Incomplete compression (IVC).
- Gas exchange.
- Mechanical friction.

The compressed air leakage into the crankcase is comparably high for the cold start at diesel engines with aluminum pistons. The clearance between the piston (rings) and the cylinder wall is still significant when cold, which is why the low speed in startup mode leads to a comparatively long (time-based) compression. For investigation of the cold start, the leakage has to be considered. However, it needs to be addressed in warm conditions due to a low impact compared to the enumeration mentioned above.

The difference between incomplete and imperfect combustion is defined as follows: with a lean or stoichiometric air-fuel ratio, there is sufficient oxygen to burn all the fuel with the chemical reactions:

$$H_2 + \frac{1}{2} O_2 \rightarrow H_2O \qquad \text{Eq. 2.19}$$

$$C + O_2 \rightarrow CO_2 \qquad \text{Eq. 2.20}$$

As the equations show, for an air-fuel ratio of less than one, incomplete oxygenation occurs:

- <u>Incomplete</u> combustion occurs only in sub-stoichiometric conditions.

Moreover, in the real engine, the chemical reactions can be interrupted at any state, and the conversion into thermal energy terminates imperfect:

- Lack of mixing formation or slow chemical reactions occur in <u>imperfect</u> combustion. [79]

The idealized thermodynamic cycle, which has the highest efficiency regardless of the combustion principle, is the Carnot process using two isothermal and two adiabatic (reversible) changes of state. This comparative process comprises these four thermodynamic changes of state, whereby the high-pressure and low-pressure loops are relevant to the overall process. However, the comparison process mainly considers the high-pressure loop, and advantages or disadvantages due to different valve timing and gas exchange losses are not considered. The enclosed area of a clockwise cycle process can be defined in the *pV* diagram as the work achieved. The larger this area becomes, the greater the work gained. Depending on the conversion point and the speed of the chemical energy, losses result from too early or late introduction of heat into the combustion chamber, blow-by, and wall heat transfer. This can result in higher losses due to friction or unused energy from an exceptionally late combustion phasing (little volume change work). Thermodynamically optimized combustion can thus lead to higher efficiency. However, this ideal process control cannot be fully exploited due to the time scales and the predefined piston travel. The (high-pressure) efficiency of the Carnot process mentioned above depends exclusively on the minimum and maximum process temperatures.

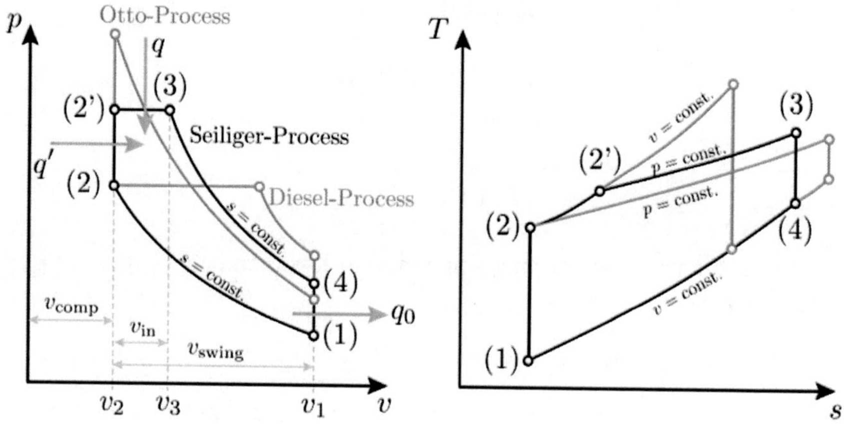

Figure 2.2: Thermodynamic Process: Seiliger in a *pV* and *Ts* Diagram [4].

The comparative process of Carnot, which considers neither irreversibility nor dissipation losses, is replaced by a more realistic Seiliger process shown in **Figure 2.2**. This process simulates both the constant pressure (diesel) and the constant volume process (gasoline) and can therefore describe the real cyclic process much better. This Seiliger process is characterized by the following:

- (1) → (2): Isentropic (reversible adiabatic) compression.
- (2) → (2'): Isochoric heat transfer; Gasoline-like combustion.
- (2') → (3): Isobaric heat transfer; Diesel-like combustion.
- (3) → (4): Isentropic (reversible adiabatic) expansion.
- (4) → (1): Isochoric heat transfer; Substitutes the gas exchange.

Furthermore, the following assumptions are made:

- The medium is thereby still considered as an ideal gas, and pressure or temperature changes do not affect the gas properties.
- The compression ratio is assumed directly from the real engine.
- No overflow or purge losses occur.
- The system boundaries are adiabatic; Blow-By = 0.
- All walls are adiabatic; Wall heat transfer = 0.

For PCCI combustion, the state changes of the Seiliger process for gasoline and diesel converge even more into each other since the objective is to reduce diffusive combustion. The premix fraction is approximately proportional to the increase in constant volume combustion. This is due to the early combustion position (compared to conventional diesel combustion) and the low diffusive fraction. On the one hand, the absence of diffusive combustion reduces the soot formation in the late combustion phase. On the other hand, it increases the peak pressure of the combustion (compare conventional and PCCI peak pressure in **Figure 5.1**).

For no kind of combustion is there such a precisely defined thermodynamic change of state as in **Figure 2.2**. The difference between theory and practice is the real losses, which is why the individual phases cannot be concretely separated from one another. Instead, a theoretical consideration of the cycle must be made section by section. Weberbauer et al. have proposed a "generally valid loss division" – called "allgemein gültige Verlustteilung" – for this purpose, by which the individual losses can be delimited from the thermodynamic optimum [77].

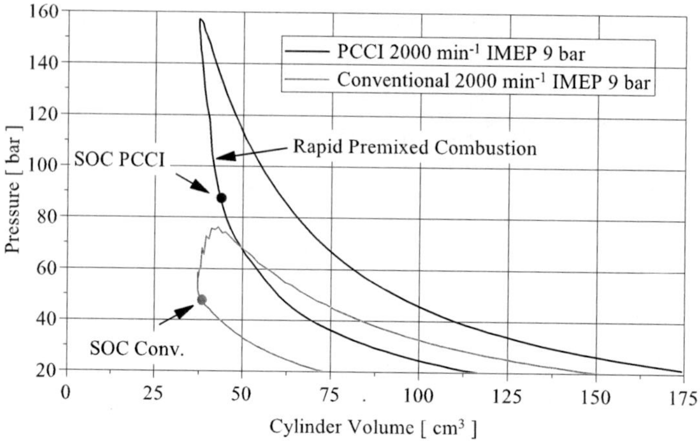

Figure 2.3: Thermodynamic Process: Conventional and PCCI Combustion (2000 min^{-1} and IMEP 9 bar, Best Points as in Section 4.5) in a *pV* Diagram.

Figure 2.3 is intended to illustrate real combustion in the medium load range. Here, conventional and partially premixed diesel combustion are plotted in a pV diagram. **Figure 4.11** and **Figure 4.15**, among others, show these two measurement points in relation to each other in terms of their emissions and combustion noise. In this figure, a comparison is to be made with the Seiliger process from **Figure 2.2**. For conventional diesel combustion in the medium load range, no consideration needs to be given to the peak pressure of the combustion (engine up to 175 bar), which is why the combustion is operated either at optimum specific fuel consumption or with regard to the exhaust gas components. In the case of consumption-optimal operation, the result is roughly as shown. However, contrary to the Seiliger or pressure-constant process, this is to be regarded as real. The conversion occurs rapidly due to the premixed portion of the first pre-injection. Further combustion ends significantly after the maximum pressure at around 222 °CA. A different combustion pattern emerges for PCCI combustion despite the same speed and load. The very early fuel conversion before the top dead center leads to a significant pressure rise up to about 157 bar at 183 °CA (shortly after TDC). Low-temperature combustion begins at 155 °CA, the NTC area begins at 159 °CA, and high temperature ignition at 163 °CA. After that, fuel is injected via main injection into the ongoing high temperature combustion of the pre-injections. This oxidization begins after a short ignition delay, resulting in an injection-dependent heat release.

Overall, PCCI combustion occurs significantly earlier, leading to a thermodynamic disadvantage. However, the burn-through rate of PCCI is 25 percent (5 to 95 %MFB) faster than that of conventional diesel. Therefore, it is difficult to directly compare partially premixed diesel combustion with a pre-defined thermodynamic process. Furthermore, the conditions in the cylinder are different for each load and engine speed point, which is why only the following statements can be made:

- An increase in thermodynamic efficiency would result from a prolonged ignition delay, whereby the start of combustion and, thus, the phasing of combustion could shifted a several degrees crank angle after TDC.
- The early and rapid combustion has the advantages of the constant-volume process. Theoretically, this thermodynamic efficiency is several percentage points higher than the pressure-constant process. Furthermore, the high compression ratio improves the efficiency of the constant-volume process compared to a conventional gasoline engine.

2.2 Rate of Heat Release and Burn Rate to Characterize the Combustion Process

The combustion can be described using various parameters. For example, the pressure gradient is important for partially premixed diesel combustion, which directly influences combustion noise and the mechanical stress of the engine. In addition to the combustion parameters, which describe the combustion as a single value, the heat release over time is a widely used evaluation criterion for combustion. In the initial phase of development using indication systems, this was calculated from the deviation between the unfired and the fired pressure curves.

The single zone models currently in use apply the 1st law of thermodynamics to describe the rate of heat release (see Eq. 2.9). This allows an energetic global consideration. However, the combustion process is not divided into local zones, so a description of local phenomena is impossible. A more detailed consideration of the combustion chamber can be made using two-zone or multi-zone models. These usually divide the cylinder mass into an unburnt and a burnt portion, in which different properties are assumed and calculated. Furthermore, quasi-dimensional models are valid for describing local effects. These are most necessary in zero-dimensional modeling to represent local phenomena. The calculation based on measured data (crank angle and cylinder pressure) allows an accurate representation of the energy conversion process for various combustion processes. However, with the charge temperature, pressure at IVC, and the calorific value, the resulting ROHR can be calculated directly using Eq. 2.9. To calculate the burn rate from the ROHR, some assumptions must be made. These assumptions are described in section 2.1. As an example, the wall heat transfer can be mentioned. This can be estimated very well using various approaches based on empirical and analytic assumptions, but it also can be misinterpreted in the case of different combustion processes. For example, homogeneous auto-ignition and diffusive combustion can be mentioned here. In the case of homogeneous combustion, the flame reaches almost the combustion chamber wall before it expires, so a higher heat input in the cylinder wall has to be expected. On the other hand, the local energy input is very decisive for diffusive combustion. This results in different wall heat transfers, which are not considered individually in this project. As already mentioned, these losses must be calculated as accurately as possible for the comparison of pressure curve analysis and work process calculation.

Eq. 2.8 can be rearranged (see Eq. 2.21) to calculate the fuel conversion efficiency. This efficiency can theoretically also be calculated by calculating the exhaust gas composition:

$$\eta_e = \frac{Q_B}{m_f \cdot H_u} \qquad \text{Eq. 2.21}$$

More adequate to assess the combustion of PCCI in the first step is by using the ROHR. For this purpose, the integration of the ROHR consists of the converted fuel mass minus the wall heat losses and leakages. Considering the internal energy without considering taking pressure and gas composition leads to Eq. 2.9. Transition from differentials to differences yields:

$$\frac{\Delta Q_h}{\Delta \varphi} = \frac{\frac{\kappa}{\kappa - 1} \cdot p_c \cdot \Delta \mathrm{V} + \frac{1}{\kappa - 1} \cdot V \cdot \Delta p_c}{\Delta \varphi} \qquad \text{Eq. 2.22}$$

And a linear approach for the temperature dependency of $\frac{1}{\kappa - 1}$:

$$\frac{1}{\kappa - 1} = 2.39 + 0.0008 \cdot \frac{V_{120}}{p_{120} \cdot V_{120}} \cdot p \cdot V \qquad \text{Eq. 2.23}$$

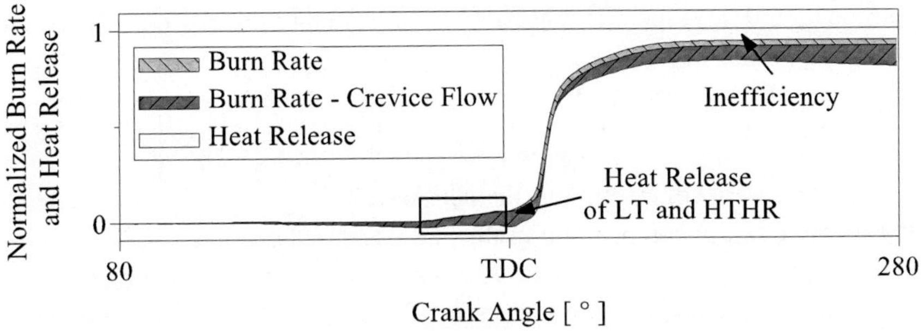

Figure 2.4: Comparison of Cumulated Heat Release, Burn Rate and Crevice Flow for a PCCI Strategy.

Figure 2.4 shows the deviance from cumulative ROHR to the cumulative burn rate. The normalized y-axis shows at "1" the possible energy conversion by the introduced fuel. The losses between this line and the burn rate are due to incomplete combustion. This efficiency can be calculated using Eq. 2.21. Another loss occurs by wall heat transfer, most likely around the top dead center. Losses also occur due to crevice flow, which is most significant in the high-pressure range. Thus, there are occasionally significant losses between the fuel energy introduced and the calculated ROHR.

2.3 Premixed Combustion: The Positive and Negative Influence on Emission

Reciprocating engine combustion has disadvantages in emissions due to the time per cycle, wall impingement, exhaust aftertreatment conditioning, and others. For ideal stoichiometric combustion, it follows that fuel oxidizes to carbon dioxide and water in the presence of oxygen:

$$C_xH_y + \left(x + \frac{y}{4}\right) \cdot O_2 \Rightarrow x \cdot CO_2 + \frac{y}{2} \cdot H_2O \qquad \text{Eq. 2.24}$$

Carbon dioxide and specific fuel consumption are thus directly correlated, and the fuel's carbon-to-hydrogen ratio and engine fuel efficiency are also relevant to reducing CO_2 emissions. In the aforementioned fuels report, the carbon-to-hydrogen ratio is 5.83.

Low-temperature combustion is characterized by chemical kinetics. Depending on the mixture composition in the end gas, combustion can be very slow and imperfect, or it can occur in simultaneous volume ignition with a very rapid, uncontrollable increase in pressure. The high exhaust gas recirculation rate contributes to a delay in the chemical reactions. As a result, the ignition limit is reached for low load and speed, and the combustion air-fuel ratio limit is reached for higher load and speed.

Partially premixed diesel combustion has negative effects that are known from gasoline engine combustion:

- Combustion is imperfect and occurs in relatively high HC and CO emissions.
- A stochastic dependency of cycle-to-cycle variations due to strong pressure rise gradients leads to noise and can damage the engine (as knocking in a gasoline engine).
- Charge stratification is only incomplete and leads to cycle-to-cycle fluctuations during combustion.
- Near stoichiometric combustion due to a high amount of recirculated exhaust gas. Local rich areas lead to incomplete combustion.
- Cycle-to-cycle variations by partially or homogeneous mixture due to sensitivity on interferences such as mixture formation, EGR rate, thermodynamic conditions, and air flow [54].

The reduction of both soot and nitrogen oxide is reached by:

- Partially or near homogeneous combustion (early main injection with EOI before SOC).
- Suppression of diffusive combustion.
- Decrease of local flame temperature: lowered nitrogen oxide formation and reduced soot oxidation - suppression of soot formation with adequate homogenization.
- Recirculation of internal and external exhaust gas: internal recirculation is needed to achieve the quantity, but is not ideal since the high temperature shortens the ignition delay of the pre-injections. However, a positive effect can still be observed as a result of internal recirculation.

2.4 Gas Exchange Calculation

A wide range of engine types is available on the market, and different types have been developed over time. They start from 2-stroke reciprocating engines with high power or small "position-independent" types to 4-stroke small to large-size engines. Other engine types as the Wankel engine [80] with an eccentric rotating design for the rotor or free-piston engines, also work in a thermodynamic open loop cycle. Besides reciprocating and rotary engines, continuous combustion in gas turbines or ramjets also operates in an open cycle. The gas exchange is – as the combustion – continuous. An oxidizer is required to burn the injected fuel for all engine types where fuel (or similar) is used. The exchange of this oxidizer and fuel with the burnt mass during the gas exchange cycle is time critical and responsible for the engine's behavior. The strategies for 2-stroke engines can be piston-controlled ports or uniflow scavenging. The drivetrain design variety for 4-stroke reciprocating engines reaches from a single camshaft in the crankcase to multiple overhead camshafts. The number and arrangement of valves depends on the engine's application, legislation, performance, and other factors. Nowadays, passenger cars commonly use four valves per cylinder– two for the intake and two for the exhaust gases. There are more disadvantages with two, three, and five valves per cylinder. Nevertheless, the degree of variability in the valve train to optimize the gas exchange for different loads and speeds, the tumble or swirl [81, 82] to improve combustion by supporting the mixture, the internal exhaust gas recirculation to lower NO_X or to improve the pumping losses, respectively, is immense. One main reason for valve trains used today is cost, but other considerations include emissions legislation, drivability, and engine power output. [83] As the system's complexity increases, the engine unit costs rise, but the benefits for engine performance (power, emissions, drivability) are not negligible.

Figure 2.5 shows the Mercedes Benz OM642 cylinder head's valve lift and flow coefficient. The crank angle designated BDC and TDC represents the gas exchange's top and bottom dead center. The flow test bench at the IFS of the University of Stuttgart is variable with respect to the flow direction. Since both flow directions are necessary for calculating the gas exchange, both curves were measured up to 10 mm valve lift (for intake and exhaust).

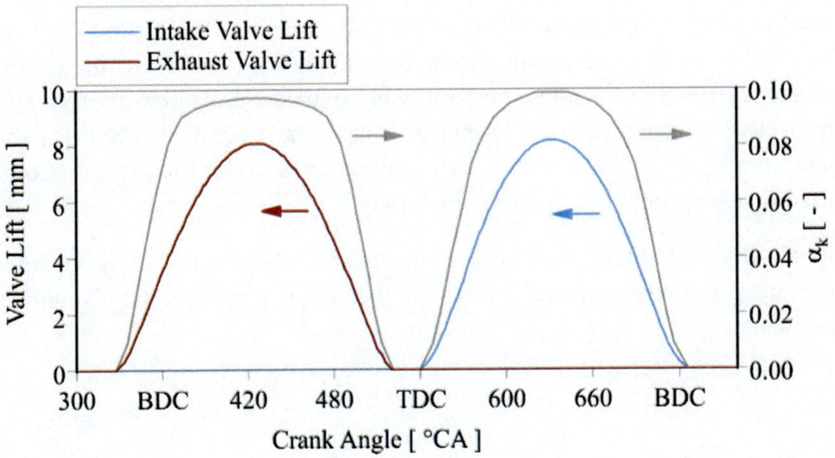

Figure 2.5: Mercedes Benz OM642 Valve Lift and Flow Coefficient (α_k). Valve Timing and Flow Coefficient Measured at the Flow Bench. Red: Exhaust Valve Lift; Blue: Intake Valve Lift [84].

In the figure, however, only one flow direction coefficient is shown in each case. For the outlet valve, the blowing direction is shown – for the flow coefficient curve of the intake valve, the test bench was used in suction mode. The flow coefficient (as α_k, also known as c_v and c_f) represents the ratio of the isentropic to the geometric cross-sectional area:

$$\alpha_k = \frac{A_{is}}{A_P} \quad \text{Eq. 2.25}$$

With the surface area of the piston:

$$A_P = \frac{\pi \cdot d_P^2}{4} \quad \text{Eq. 2.26}$$

The software TIGER from EnginOS GmbH was used for the gas exchange analysis. For the equations in section 5, the internal EGR rate calculation results were used.

Furthermore, static measurements of the flow characteristics for the valves of an internal combustion engine can be used to obtain good results for gas exchange calculations. Another possibility is a dynamic measurement. Here, the respective camshaft (exhaust or intake) is driven at a constant speed. The dynamic valve lift curve is then determined via distance sensors. It can also be used to determine the flow coefficient. [85]

Four-valve technology with bucket tappets and two camshafts (one each for the intake and exhaust valves) is used for the used engine at the single-cylinder test bench.

3 Single-Cylinder Test Bench and Measurement Equipment

The test bench at the IFS (University of Stuttgart) is fully equipped with an external compressor, the possibility of independent conditioning of air, water and oil temperature as well as intake-air, and oil pressure. A cooled high-pressure EGR line with an EGR valve to apportion fresh air and recirculated air is traced back to the intake plenum with the water jacket. Furthermore, the downstream piping is designed to reach an equally premixed mixture of fresh and recirculated air. The intake plenum originates from the six-cylinder V engine, so the plenum size applies to all three cylinders of this cylinder head. The cylinder head is also an adaptation of the full-size engine for single-cylinder use, but cylinders two and three are used as dummies. An entire camshaft is inserted, but there are no valves for cylinders two and three. Furthermore, the CO_2 content in the intake plenum is measured to calculate recirculated exhaust gas rate. The measuring point is inserted in the intake plenum on the opposite side of the intake ports and valves, and the equation to calculate the EGR rate is as follows (based on [86, 87]):

$$\frac{EGR}{[\%]} = 100 \cdot \frac{\frac{CO_{2,Inlet}}{[\%]} - \frac{CO_{2,Atm}}{[\%]}}{\frac{CO_{2,Exhaust}}{[\%]} - \frac{CO_{2,Exh}}{[\%]}} \qquad \text{Eq. 3.1}$$

The CO_2 content of the intake, exhaust, and atmosphere defines the equation EGR rate (%).

The measurement point in the exhaust pipe for the Horiba 7170 DEGR is located 30 cm downstream of the exhaust valves. NO_X and THC are measured wet. After drying the exhaust gas, its components CO (high), CO (low), CO_2, and O_2 are detected. The air-fuel ratio is measured by an ETAS Lambda Meter with a Bosch LA4.2 exhaust gas oxygen sensor. The Lambda defining the air-fuel ratio with stoichiometric balancing is also calculated using the Brettschneider formula [88]. The third way to determine the air-fuel ratio is by calculating the actual air mass flow in relation to the injected fuel mass (time-

M. S. Wahl, *Emission Reduction with an Alternative Diesel Combustion Process*, Wissenschaftliche Reihe Fahrzeugtechnik Universität Stuttgart,
https://doi.org/10.1007/978-3-658-42094-9_3

dependent). While the fresh air mass flow is measured, the recirculated exhaust gas flow must be calculated.

Fault detection is feasible by comparing the Bosch LA4.2 oxygen sensor, the air-fuel ratio of the calculated Brettschneider, and determining the supplied air and fuel. Both the Brettschneider equation and the calculation of recirculated air, which is relevant to determine the cumulated mass flow of aspirated air, depends on the intake and exhaust pipe measurement of CO_2. Therefore, a CO_2 intake measurement fault could be missed by only one measuring device, but in combination, a system error can be detected in most cases.

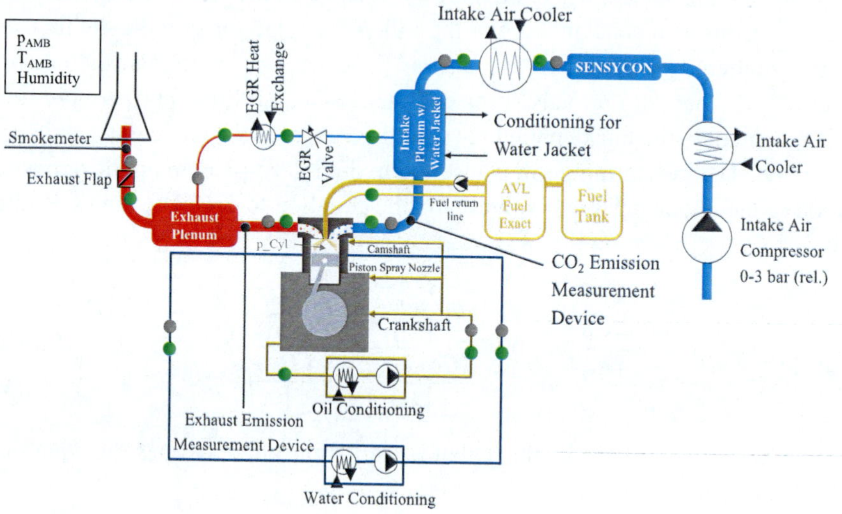

Figure 3.1: Single-Cylinder Test Bench Configuration with Measurement Equipment. Green Dots for Temperature, Gray Dots for Pressure Measurement Points. Adapted from [2].

Figure 3.1 shows further specifications and measuring points of the test bench setup. The temperature and pressure measurements are shown as green and gray dots, respectively. No distinction is made in the figure between the measurement principles such as type N or K for thermocouples, PT100, PT1000, or piezoelectric/piezoresistive and constant pressure sensors. The conditioning units shown in the picture are each capable of heating and cooling. A heating element and a heat exchanger for cooling are installed in each case to condition the medium. If the flow rate is low, a combination of heating and cooling is recommended in order to be able to set a stable temperature. As cooling fluid for the heat exchanger, a cold-water circuit from the building services is available for all test benches.

Table 3.1 gives a list of the single-cylinder test bench configuration is given. Section 4.1, especially **Table 4.1**, provides additional information on the engine and injection system:

Table 3.1: Test Bench Configuration.

Test Bench	Single-Cylinder Only
Electrical Dynamometer	Siemens 1GP2219
High Pressure Sensor	Kistler 6041A
Low Pressure Sensor (Intake)	Kistler 4045A5V39
Low Pressure Sensor (Exhaust)	Kistler 4075A10V39
Switching Adapter	Kistler for exhaust pressure
Exhaust Gas Measurement	Horiba 7170 DEGR
AVL Smoke Meter	415S G002
ETAS Lambda Meter	LA4 E
Fuel measurement	AVL FUEL EXACT PLU300 FF
Fuel Type	Diesel
Fuel Density	829.1 kg/m^3
Fuel Cetane Number	51.9
Lower Heating Value (Net Calorific Value)	42.918 MJ/kg
Fuel Pressure (investigated)	40 to 140 MPa
Intake Air Charge (Pressure)	External Compressor
Indication System	IndiGO V10.58
Test Bench Automation	TECHNOGERMA Systems GmbH Autotest2000 Pro 4.7
Oil Temp. @all measurements	90 °C
Cool. Water Temp. @all measurements	90 °C
Cool. Water Temp. for pressure sensor	50 °C

3.1 Particulate Matter Measurement

In addition, an AVL Smoke Meter 415S measures the blackening of filter paper at exhaust backpressure up to 0.75 bar relative with a special extraction according to ISO 10054 [89–91]. The Smoke Meter uses an electronically controlled, continuously aspirating diaphragm pump with flow metering instead of the piston pump of the earlier measuring devices. This results in constant exhaust gas extraction and, thus, better repeatability. In ISO 10054, PM, also called soot, is described as “all components contained in the exhaust gas which blacken the filter”. The AVL 415S was developed to measure relatively low concentrations relevant to this project [33]. The measuring point is located behind the exhaust flaps since the maximum permissible absolute pressure of 1.75 bar upstream of the exhaust flaps is exceeded in some measuring points. The sampling probe is inserted into the exhaust pipe as described in the instructions – ascending in a straight piece of pipe, with six times the pipe diameter upstream or downstream no curve is present. In conjunction with the backpressure valves, the exhaust plenum results in comparably low-pressure pulsations of the exhaust gas near the sampling point. The filter paper used in the 415S is Viton instead of silicone, as in the 415. Furthermore, the heating of the 415S allows more accurate results with increasing sample volume (sampling time). The main reason is the loss of condensate at increased suction volume on the surface of the suction tube and in the instrument itself. This results in reduced blackening and can cause the wall deposits to detach during a subsequent measurement, leading to an increased, incorrect measurement result. The Smoke Meters output is the Bosch Filter Smoke Number (FSN), where the range of FSN is between zero and ten. [89]. These thresholds can be derived from Eq. 3.2 and Eq. 3.3. For zero, the ratio of $R_b^{'}$ and $R_c^{'}$ in Eq. 3.3 is one – which means that the clean filter and the used filter paper have the same optical reflectance – inserting this in Eq. 3.2 leads to an FSN of zero. On the other side, for an FSN of ten, the $R_b^{'}$ to $R_c^{'}$ ratio in Eq. 3.3 is infinite, and therefore the FSN in Eq. 3.2 ten.

According to ISO 10054, the FSN is calculated as follows and depends on the relative reflectometer value $R_r^{'}$:

$$FSN = \left(1 - \frac{R_r'}{100}\right) \cdot 10 \qquad \text{Eq. 3.2}$$

where $R_r^{'}$ is:

$$R_r' = \left(\frac{R_b'}{R_c'}\right) \cdot 100 \qquad \text{Eq. 3.3}$$

Therefore, $R_b^{'}$ is the measured value of the blackened filter, and $R_c^{'}$ that of the clean filter, respectively.

Various methods for converting the FSN to an equivalent soot mass are described in the literature. For example, an empirical-based calculation to convert the Bosch number (as a correlation of Hartridge and Dunedin smoke meters) was published as early as 1965 by the Motor Industry Research Association (MIRA) [92]. Unfortunately, this empirical approach's lowest recorded smoke number is 1.0, and the extrapolation accuracy cannot be predicted. Therefore, this is not suitable for modern diesel engines with LTHR regimes since they operate with FSN below 1.0 in most operating points and conditions. Other approaches to transfer the FSN to mg/m^3 by Alkidas [93], Muntean [94], and Christian [95] are compared in [96]. Christian's correlation is the most promising, with a slope of 0.99 and an R^2 of 0.94 for all conditions, a good correlation.

However, soot formation is complex due to the interrelated mechanisms and effects of thermo- and fluid dynamics [54]. The representation and transferability of different measurement systems are difficult caused by the fact of different emitted particles (e.g., size, composition, aggregate state). For this reason, the filter measurement method was used in this project, as it provides very robust measuring results. Transferability to results in other publications is possible due to the standardized measurement setup, but only the self-generated measurement results with the constant measurement setup are compared in the following chapters.

3.2 Nitrogen Oxide Measurement

Nitrogen Oxides (NO_X) – mainly summarizing nitrogen oxide (NO) and nitrogen dioxide (NO_2) – are formed via various mechanisms in the combustion process. The main source of nitrogen and dioxide for these mechanisms is fresh intake air. Less relevant and, therefore, not described here are N_2O, N_2O_3, and N_2O_5. A schematic for the formation of NO (and NO_2) in a DI-diesel engine spray is shown in **Figure 1.3**. For the test bench using a HORIBA MEXA 7170DEGR, the two-point calibration range for NO is zero and 1013 ppm, as additionally listed in **Table 3.2**. In addition, a test gas at 96 ppm was used periodically to check the measurement accuracy in lower values. This deviation was always less than 1 %.

Prompt NO; Fenimore: The formation of prompt NO, also known by its inventor Fenimore, contributes only a small part of the total NO_X emission in the internal combustion engine. The reaction proceeds rapidly via the formation of hydrogen cyanide. This requires CH radicals, which is why the formation of prompt NO basically occurs in the flame front. The activation energy is lower than that for thermal NO. The starting temperature of the Fenimore reaction is about 1000 K.

Thermal NO; Zeldovich: A primary process of nitrogen oxide formation in the piston engine due to the Zeldovich mechanism depends on the local combustion temperature, pressure, and air-fuel ratio. The activation temperature for the start of the Zeldovich mechanism is given in the literature as 1700-2000 K. [97, 98]. Essentially, the following elementary reactions are decisive:

$$N_2 + O \Leftrightarrow NO + N \qquad \text{Eq. 3.4}$$

$$O_2 + N \Leftrightarrow NO + O \qquad \text{Eq. 3.5}$$

$$OH + N \rightarrow NO + H \qquad \text{Eq. 3.6}$$

Moreover, the Zeldovich reaction scheme was extended and super-extended by Miller et al. in 1997 and 1998 [99]. The high activation temperature and the activation energy (around 316 kJ/mol for equation Eq. 3.4) are based on the chemical stability of N_2 due to its triple bond. The formation of N determines the reaction rate of equations Eq. 3.5 and Eq. 3.6. Since the adiabatic flame temperature is about 2500 K, the Zeldovich reactions start slightly behind the flame front in oxygen-rich regions where a high amount of NO occurs.

The most significant reaction of NO_2 is:

$$NO + HO_2 \Leftrightarrow NO_2 + OH \qquad \text{Eq. 3.7}$$

The chemical equilibrium at normal temperature is on the right side (NO_2). Thus, NO reacts with sunlight after a few hours or days. [21]

Fuel-N_2 conversion: The formation/conversion of nitrogen oxides via the nitrogen in the fuel is not relevant for internal combustion engines. The amount of N_2 in liquid fuel – in contrast to solid fuel such as coal – is almost zero.

Nitrous oxide: The formation of NO via nitrous oxide is characterized by high pressure and comparatively low-temperature. In addition, the local air-fuel ratio must be lean since the reaction occurs mainly in the presence of excess oxygen. At higher temperatures, the pre-described Zeldovich mechanism then dominates. When operating with partially premixed combustion, this formation path for NO is distinctly possible since the combustion phasing causes the in-cylinder pressure to rise sharply even at partial load.

It is conceivable to influence the local air-fuel ratio behind the flame front or to lower the temperature and pressure to reduce the formation tendency. In order to fulfill the air-fuel mixing behind the flame front (where the NO formation takes place), homogenization before the SOC gives satisfactory results. With the temperature and pressure reduction, the peak pressure of combustion should be shifted to the expansion phase. Therefore, the retardation of injection timing is used (see **Figure 1.2**), but the increasing soot formation leads to the described trade-off between soot and nitrogen oxides.

Besides the variations of injection pressure, injection timing, and air-fuel ratio in and behind the flame front, the global reduction of oxygen in the combustion chamber effectively reduces nitrogen oxides. However, reducing the oxygen content also has a side effect: the high heat capacity of the inert gas. As long as the inert gas is CO_2, exhaust gas (internal or external recirculated), or N_2, a temperature reduction in the flame front will reduce the Zeldovich formation mechanism. At the relevant diesel combustion temperatures, the heat capacity per mole of CO_2 is almost twice that of N_2 – that of exhaust gas is slightly higher than that of N_2 [19].

3.3 Additional Pollutant Measurement

The ideal chemical equilibrium in the combustion of pure fuel and air mainly results in the non-toxic exhaust gas components: nitrogen (N_2), steam (H_2O), carbon dioxide (CO_2), and, in lean-burn engines, additional oxygen (O_2). This type of complete combustion is achieved under laboratory-optimized conditions but rarely in an internal combustion engine. The chemical reactions in this process are time-dependent and may not be in complete chemical equilibrium.

Total hydrocarbons (THC) or hydrocarbon (HC) emissions result largely from the liquid fuel used, which is based on hydrocarbon components. The HC component in the exhaust is therefore based on unburned fuel. Therefore, optimizing combustion quality to reduce engine-out HC emissions is important. This optimization can be achieved, for example, by adjusting the combustion chamber geometry, the airflow, or the injection strategy, as this positively influences the local composition of parameters such as the air-fuel ratio, fuel vapor, turbulence, and heat with regard to optimized combustion.

The presence of non-oxidized CO in raw emissions is similar to the existing HC. Incomplete combustion due to lack of oxygen in the combusting area interrupts the chemical reaction with O_2 to CO_2.

The virtual absence of SO_X (accumulation of SO_2 and SO_3) in the exhaust gas for on-road specifications is attributable to the legislation and regulation of the fuel. In 2001, the limit of sulfur in diesel was lowered to 50 ppm. Two years later, a level of 10 ppm was introduced and labeled as "Sulfur oxide free". [100] The diesel used for this project has a sulfur content of 8.4 mg/kg according to ISO 20846:2011.

The analyzer rack (ANR) of the Horiba MEXA-7000 engine exhaust analyzer used contains seven slots for ultra-compact analyzers. An independent configuration to specify each ANR for direct or (constant volume sampling) CVS diluted samples is included. In addition, the main control unit (MCU), interface controller (IC), solenoid valve selector (SVS), sample handling system (SHS), and power supply unit (PSU) are housed in the MEXA-7170DEGR rack. This rack is located in the control room to minimize pressure and temperature fluctuations. Also next to the test bench is an oven (OVN), also known as the heater analysis unit. This oven houses the high-temperature analyzers for the detection of THC and NO/NO_X (measuring principle in **Table 3.2**). The length of the heated sampling line from the engine to the OVN is limited in all cases. The integrated sampling pump supplies all instruments (OVN and MEXA rack) with the exhaust gas. [101, 102]

Table 3.2: Measured Emission Components on the Test Bench with a HORIBA MEXA 7170DEGR.

Emission Component	Abbreviation	Calibration Range	Calibration Gas	Measuring Principle
Carbon Monoxide Low	CO_{Low}	0–925 ppm	CO	NDIR
Carbon Monoxide High	CO_{High}	0–4.51 Vol%	CO	NDIR
Carbon Monoxide	CO_2	0–14.94 Vol%	CO_2	NDIR
Oxygen	O_2	0–4.71 Vol%	O_2	MPA
Nitrogen Oxide	NO_{High} / NO_X	0–1013 ppm	NO	CLD
Total Hydrocarbon	THC	0–1484 ppm	C_3H_8	FID

To reduce tailpipe emissions of THC, CO, and NO_X (as described in section 3.2), a three-way catalyst can be used for stoichiometric engine applications. The conversion rate for a heated catalyst under stoichiometric conditions can exceed 98 %. [103] For lean combustion strategies, such as in a diesel engine, a diesel oxidation catalyst (DOC) is used to reduce both HC and CO. The DOC also reduces particulate matter – soluble organic fraction (SOF) – by oxidation of long-chain hydrocarbons. In addition, NO oxidizes to NO_2. The equations for reduced chemistry of the DOC are Eq. 3.8 to Eq. 3.10 [54, 104].

$$2CO + O_2 \rightarrow 2CO_2 \qquad \text{Eq. 3.8}$$

$$C_xH_y + \left(x + \frac{y}{4}\right) O_2 \rightarrow xCO_2 + \left(\frac{y}{2}\right) H_2 0 \qquad \text{Eq. 3.9}$$

$$2NO + O_2 \rightarrow 2NO_2 \qquad \text{Eq. 3.10}$$

3.4 Injection System Measurement

Moreover, various diesel injection systems are used today that were developed in recent years, including pressure build-up (pump), storage (rail), type of injector, and position of the injector. Good literature on the history of fuel injection published by Frank DeLuca can be found in [105]. With respect to pressure build-up by combustion, a division between globalization and localization is key. While in local systems without storage, the high-pressure build-up occurs in the injector, a global system uses a high-pressure pump and rail to supply all injectors. These injectors are activated either by fuel pressure or an external electrical signal. The principle of needle lifts in externally activated (electrical signal) injectors is based – with a few exceptions – on piezoelectric or solenoid valves. The advantages of pressure-activated injectors arise in terms of needle lifts per cycle and the variability of injection pressure, duration, and timing. In summary, a combined high-pressure pump injector system has variability disadvantages, unlike a pump-rail injector system where the most variable system has a piezoelectric activated valve lift. This decoupled system, also known as the common rail system, has a wide range of injection pressure, injections per cycle (number and timing), needle lift, fuel quantity per injection, and injection timing (not camshaft dependent).

In contrast to mechanical injection systems with camshaft actuation, the electrical injectors have further advantages with respect to external influences such as fuel temperature, variations in exhaust aftertreatment (e.g., active regeneration of the soot filter system), and droplet size [106]. For these reasons, electrically actuated systems offer advantages for current and future emission regulations. In addition, the maximum injection pressure of pressure-actuated systems is comparatively low (about 200 bar), and the combustion chamber geometry is therefore formed in a pre or swirl-chamber (IDI). However, this leads to efficiency losses due to the airflow into the pre or swirl-chamber.

The use of modern injection systems as the main responsible for reducing raw emissions requires accurate calibration of the overall system and the injector. This involves not only pre-calibration to compensate for variations in the manufacturing process but also constant adjustment of the injection volume and rate shaping due to deposits and wear of the injector. Therefore, the implementation of control and regulation mechanisms such as zero quantity calibration (described in [107] and in [54]), pressure wave compensation, injection quantity adjustment (IQA), and average quantity adaption continuously check the quantity and quality of the injection system.

In addition, fuel pressure control for the high-pressure rail can be performed on the high-pressure or suction side of the fuel pump. The combination of high-pressure and suction-side control also combines the advantages of individual control systems.

Recent publications show another way to reduce engine-out emissions influenced by a better mixing process satisfactorily. For this purpose, the injection system described above is used and extended by a mechanical duct to concentrate the spray jets in order to achieve better premixing with a smaller Sauter Mean Diameter (SMD) for the fuel droplets [108]. Furthermore, the piston geometry is modified so that there is no collision of the moving mechanical parts with the duct and. Also, the wall impingement of the exiting fuel jets is as low as possible for all engine map conditions.

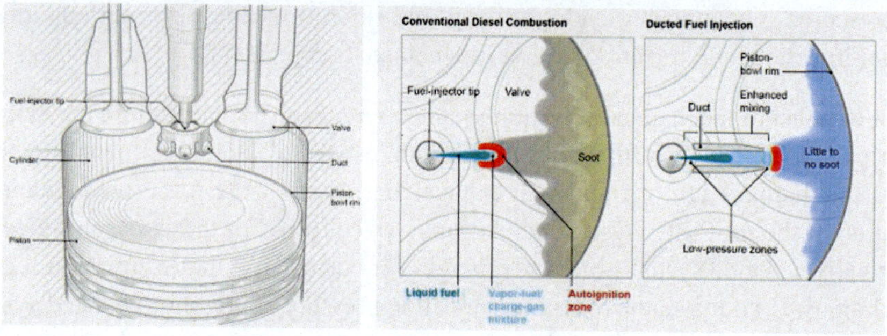

Figure 3.2: Ducted Fuel Injection to Reduce Soot Emissions in a Diesel Engine. Left: Sectional Image of a Combustion Chamber with the Duct. Middle: Schematic of Combustion/Soot Formation with Diesel Injection and Diffusive Burning. Right: Schematic of Combustion/Soot Formation with Duct at the Injectors Tip [109].

In this project, a common rail system was adapted for a single-cylinder engine. The high-pressure pump is driven externally, and the in-rail fuel pressure valve regulates the deduction quantity. The injector is a piezo-actuated system, offering a high degree of freedom regarding the injection strategy described. This injector is activated electrically by a positive voltage signal and deactivated by a negative voltage signal (see **Figure 3.3**). Activation means that the deformation of the piezo stack leads to a needle lift via a transmission ratio and a subsequent injection of fuel. In the following, the deactivation by a negative signal leads to the end of the injection.

For the injection quantity, especially for the small amount of pre-injected fuel in PCCI, it is relevant to use an injector with sufficiently high repeatability. While a solenoid valve injector has disadvantages in the ballistic range for minimum injection quantities, piezo injectors have been developed with respect to near-zero shot-to-shot deviation and a maximum of eight injections per cycle [54]. The high displacement force and short switching times allow multiple high-precision injections per cycle. Disadvantages in serial use are the high initial costs for the injector itself and the ECU's electrical control. [110, 111]

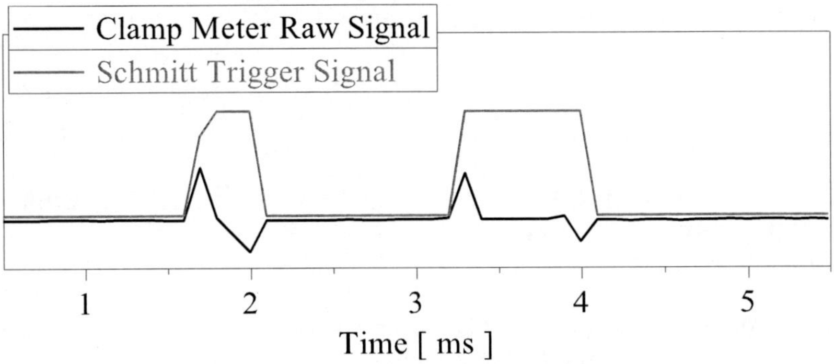

Figure 3.3: Measured and Schmitt-Trigger-Optimized Clamp Meter Signal of Two Injections. Black: Clamp Meters Raw Signal. Gray: Schmitt-Trigger Signal (Adapted from [54, 111]).

The electrical actuation signal from the ECU to open and close the piezo-inline injector to inject is shown in **Figure 3.3**. A clamp meter to measure the load current is installed to pick up the signal of the ECU to the injector. The black line represents the raw electrical signal in **Figure 3.3**. When opening and closing, a positive and negative area can be identified in each case (see figure); this is recorded stably despite the resolution of a one-degree crank angle. However, the signal strength (rise) varies depending on the sampling time. Therefore, no clear injector control can be represented from this signal, both on the test bench and in evaluation. An electrical circuit was connected to the clamp meter's output to obtain a stable and reproducible signal. This circuit reacts to a positive and a negative threshold value. Suppose this threshold is exceeded or undercut; the output signal of the electric circuit switches to a logically high or low. High, in this case, means an output signal of 4.5 volts; 0.5 volts defines the zero state. This output signal is also shown in the figure with a gray line. The relationship between the black and gray lines can be easily seen, with the gray line's electrical signal visually representing injection. This electrical circuit is called a "Schmitt Trigger", where trigger means that the threshold value of the circuit only triggers the output. Thus, the output retains its value as long the "trigger" is not activated. The positive signal of the black line is used as the upper threshold value; the negative signal is used as the lower threshold value.

3.5 Indication System for Intake, Exhaust, and In-Cylinder Pressure Measurement

This section is intended to provide an overview of the indication system used at the test bench. Two low-pressure sensors in the intake and exhaust pipe as one high-pressure sensor in the combustion chamber is used. The signals from these three pressure sensors are further processed in Kistler charge amplifiers Type 5064 and piezoresistive amplifiers Type 4665, respectively. Finally, the Kistler signal conditioning platform (SCP – Type 2853) is used for post-processing the signals.

The water-cooled high-pressure sensor is screwed lateral into an M8 borehole on the side of the exhaust valves. The front of the pressure transducer is recess mounted (not flush with the combustion surface); therefore, whistle-vibrations are possible depending on the inflow area [112]. Furthermore, a continuous cooling water flow without air residual is relevant to reach a good quality measuring result since the piezoelectrical measurement principle is highly sensitive to influences.

A switching adapter for the low-pressure sensor in the exhaust pipe, a switching adapter prevents the piezo-crystal from degrading at high temperatures (Curie temperature). It is necessary to pre-heat the switching adapter and execute a zero-offset correction for the signal to a barometer in the test cell (the fresh air conditioning or exhaust extraction may lead to modest under or overpressure in the test cell). Additionally, the coolant temperature must be stabilized to 50 °C, and pressurization of the sensor is limited to approximately sixty seconds (pre-heating) followed by 200 measurement cycles (switching adapter opens, measuring of 200 cycles, and closing of the adapter occurs).

A Kistler 2613A crank angle encoder (rotary) is adapted on the crankshaft. An extension rod reduces housing vibrations by tightening directly with the crankcase housing. The encoder electronically processes the raw signal into a CDM and TRG signal. The split per revolution is 720 CDM and one TRG. The measuring system has no dependence on the crankshaft position during installation, which is why the position of the fired TDC must be carried out subsequently. Various methods are possible, but in this project, the TDC determination is carried out by towing the engine at operating temperature. Measuring of a few cycles (30 in this case) and averaging the results in a towed pressure curve. The maximum pressure is corrected for the thermodynamic loss angle and shifted as a function of the reference signal (TRG). Before each measuring day, the position of the TDC is checked in order to be able to detect a deviation if necessary. In case of a slight deviation, this is only noted but not updated. The TDC position must remain constant over the various measurement days; otherwise, a potential error is given on the test bench or at the engine.

- An accurate TDC prediction is relevant for further evaluation of the pressure signal. For example, with a deviation of one-degree crank angle, the burn rate delta is up to ±10 %.

Adjustin the deviation from TRG to TDC (± thermodynamic loss angle) is only performed during the first engine test. Further checks are for control purposes only.

4 Test Bench Measurements and Analysis

A state-of-the-art direct-injection diesel engine was used for the test setup. The Mercedes Benz OM642 base engine is available in many different configurations. Piston material, piston geometry, connecting rod length, and a varying compression ratio for different configurations and markets should be mentioned as the main variations. Therefore, a comparison of different research and publications with other OM642 can only be made with the knowledge of the parts used for investigation.

4.1 Single-Cylinder Mercedes Benz OM642

The single-cylinder engine used for PCCI and conventional diesel tests has the configuration as shown in **Table 4.1**.

It is common for modern diesel engines to be equipped with a direct injection system. A common rail injection system based on a Bosch CRI3-20 was used here. The piezo inline injector with eight 0.126 mm bores and an umbrella angle of 157° was originally used by Mercedes Benz in combination with the piston used in this project. Hereby, the combustion chamber is integrated into the piston bowl (as typical for standard diesel engines) and forms an omega in cross-section. While an HCCI injection strategy with a flat piston is applicable, the piston geometry for PCCI or other alternative diesel combustion strategies can vary depending on the injection timing. Therefore, state-of-the-art engine parts were used for the investigations in this paper. The injector is integrated into the cylinder head at a central position in the combustion chamber. An ETK engine control unit with the software INCA from ETAS GmbH is used for the investigations. The engine control unit allows the acquisition and storage of all parameters except the fuel pressure regulator. The most used parameters are injection timing, injection duration, and rail pressure.

In studies using alternative combustion methods, noise benefits can be achieved by modifying the in-cylinder geometry to reduce the premixed

M. S. Wahl, *Emission Reduction with an Alternative Diesel Combustion Process*, Wissenschaftliche Reihe Fahrzeugtechnik Universität Stuttgart,
https://doi.org/10.1007/978-3-658-42094-9_4

proportion at SOC. The most distinctive form of alternative combustion is HCCI, in which fuel is injected during the intake or early compression stroke. Different piston crowns can optimize mixing, reduce the surface area to lower the wall heat transfer or adjust the compression ratio. Pedersen and Schramm published investigations concerning the reduction of combustion noise. They used DME fuel (dimethyl ether) in different piston crown geometries for their studies.

Table 4.1: Single-Cylinder Mercedes Benz OM642 Engine Parameters.

Engine Type	Mercedes Benz OM642
ECU	BOSCH EDC17 CP46-1.70
Displacement Volume	498 mm^3
Stroke	92 mm
Bore	83 mm
Stroke/Rod (s/r) Ratio	0.27
Connecting Rod Length	168 mm
Compression Ratio (geo.)	15.34:1
Number of Valves	4 (one intake port with swirl flap)
Swirl Flap	Open
Injection System	CRI3-20, Robert Bosch GmbH
Nozzle Number	8 holes
Spray Angle	157°
Orifice Diameter	0.126 mm
Fuel Pressure	Up to 160 MPa
Intake Valve Opens (IVO)	5 °CAaTDC
Intake Valve Closes (IVC)	0 °CAbBDC
Intake Valve Lift (max.)	8.16 mm
Exhaust Valve Opens (EVO)	32 °CAbBDC
Exhaust Valve Closes (EVC)	18 °CAbTDC
Exhaust Valve Lift (max.)	8.1 mm
Negative Valve Overlap	13 °CA

They concluded that the diesel-type piston is preferred due to the integrated combustion chamber. Furthermore, the assumption is that two main facts reduce combustion noise: limiting exposure of the cylinder liner and the smaller diameter of the bowl. Obviously, the natural frequency differs in a smaller bowl; the generated waves do not reach the liner (wall) to transfer these waves. [113]

4.2 Introduction to the Test Bench Measurements and the Ignition Delay for PCCI

In total, 364 stationary points were measured and evaluated for conventional and PCCI combustion strategies. The study of parameters such as EGR rate, engine speed, engine load, injection strategy, exhaust and intake manifold pressure, intake temperature, and the air-fuel ratio under steady-state conditions was performed. In addition, 199 engine cycles were measured and evaluated in terms of combustion characteristics such as maximum pressure rise per degree crank angle (combustion noise), peak pressure, and thermodynamic aspects, including MFB50, ROHR, SOC, and combustion noise of the different combustion modes. In addition, exhaust gas analysis, including NO_X, CO, CO_2, THC, O_2, and soot (FSN) was used to define the measurement as valid or invalid and to assess the combustion quality.

The injection system has a pressure sensor to regulate the fuel pressure. Additionally, a clamp meter is installed at the piezoelectric injector to determine the start and end of energization through the electrical signal. This electrical signal is positive for the start of energization and negative for its end, as shown in **Figure 1.4**. A Schmitt Trigger was installed beyond the clamp meter to generate a vivid signal for graphical interpretation. When the positive threshold is exceeded, the Schmitt Trigger output signal changes from zero to one, and for the negative threshold (end of injection), from one to zero again. This is displayed in the middle part of the following graphics to show the electrical start and end of injection in normalized scale, and the belonging axis labeling is "Injection".

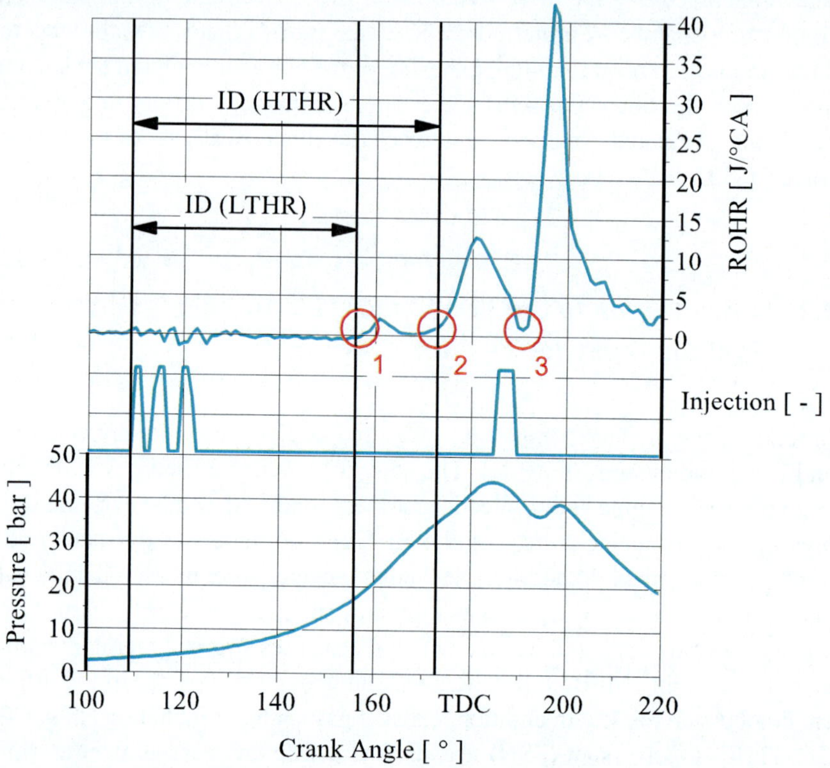

Figure 4.1: Measured In-Cylinder Pressure, Calculated Heat Release Rate and Injection Strategy of a PCCI Test Bench Measurement at 2000 min^{-1} [2].

To illustrate and describe the diagrams used to compare the strategies, **Figure 4.1** provides additional information. All diagrams show the in-cylinder pressure in the lower part, the electrical injection signal in the middle (as described above), and the rate of heat release in the upper part. For example, in **Figure 4.1**, three pre-injections and one main injection were applied. The first pre-injection starts at 108 °CAbTDC, and two other injections follow in less than 5 °CA. The time from the start of injection to SOC is labeled "ID (LTHR)" for the SOC of the LTHR and "ID (HTHR)" for the HTHR, respectively. The belonging SOC is marked in ROHR with a red circle and "1" for LTHR, "2" for HTHR, and "3" for the heat release of the main injection (in this case). The main injection's electrical signal starts at 185 °CA, and the heat dissipation can be detected at around 192 °CA. Thus, the ID is short and around 6 °CA. This heat release with three independent and separate SOCs is not common for PCCI. Mainly, the main injection starts before the HTHR starts (pre-injections), or the main injection is set while the HTHR is ongoing. In this case, it is useful to use the illustration of the ID and the two-stage ID of the pre-injected fuel.

In the case of an early single main injection without pre-injection(s), a two-stage heat release can be observed. In the case of **Figure 4.1**, the preconditioned combustion chamber exhibits high turbulence, temperature, and pressure caused by the pre-combustion. Consequently, the in-cylinder pressure and the mass mean temperature are comparably high. Thereafter, the chemical reactions of the NTC area, as described by Rether [57] and in section 1.4, are skipped. The reaction path of the HTHR (main injection) starts without an LTHR at 192 °CA.

Table 4.2: Test Parameters of the Measurements in **Figure 4.1**.

Test Parameter	Abbreviation	Unit	Value
Engine Speed	n	[min^{-1}]	2000
Engine Load	IMEP	[bar]	5
Air/Fuel Ratio	λ	[-]	1.55
MFB50 %	MFB50	[°CAaTDC]	17
Intake Pressure	p_2	[bar]	AMB
Exhaust Pressure	p_3	[bar]	1.000
EGR Rate	EGR	[%]	30
Intake Temperature	T_2	[°C]	45

4.3 Premixed Charge Compression Ignition Measurements

Partially premixed diesel combustion is known for its advantages in terms of concurrently reducing nitrogen oxides and soot emissions. The necessary knowledge for optimizing a single-cylinder diesel engine with partially premixed combustion has been built up through many parameter variations at the test bench. The theoretical basics clarify the difference between partially premixed combustion and conventional diesel combustion, with experience on the test bench leading to further knowledge.

Therefore, it is important to inject the premixed portion in such a way that the mixing processes can progress well during the ignition delay. The amount of fuel injected determines the air-fuel ratio, with a lucid difference between global and local considerations. After the pilot injections, the global lambda is comparable to conventional diesel combustion since, although a reduced quantity is introduced, it meets a large amount of recirculated exhaust gas in the combustion chamber. The mixing processes then proceed similarly. This is why a significant difference in the subsequent combustion is achieved by leaning out with the pre-injection. This means that a global lambda of two will lean out significantly less before SOC than a global lambda of five. The mixing temperature necessary for low and high temperatures and the available oxygen molecule is partly not present in very lean premix combustion, which is why ignition partly fails to occur or only burns incompletely. Whereas a mixture of lambda two or lower can even lead to very pronounced low and high temperature ignition. In this case, the present radicals trigger a chain reaction, which is extinguished much more slowly in the individual ignition spots. The disadvantage, in turn, is the high conversion rate with a richer air-fuel base mixture, which must be retarded by increasing EGR rates. However, this project solved this trade-off for load sweeps at 1500 and 2000 min^{-1}, and good results were obtained.

Table 4.3: Test Parameters of the Measurements in **Figure 4.2**.

Test Parameter	Abbreviation	Unit	Value
Engine Speed	n	[min^{-1}]	1500
Engine Load	IMEP	[bar]	4–7
Air/Fuel Ratio	λ	[-]	1.4–1.8
MFB50 %	MFB50	[°CAaTDC]	−3–−7
Intake Pressure	p_2	[bar]	1.113–1.535
Exhaust Pressure	p_3	[bar]	1.194–1.535
EGR Rate	EGR	[%]	30–50
Intake Temperature	T_2	[°C]	45

Figure 4.2 shows four operation points of partially premixed diesel combustion. These represent the best points (by emissions for load variation) at a speed of 1500 min^{-1}. In the upper part of the figure, the injection curve is also shown with a different axis representation since the pre-injections for all measuring points start at 75 °CAbTDC. It can be seen that the main injection for IMEP 7 bar had to be shifted significantly to achieve a good result. Further operating point-dependent parameters are shown in **Table 4.3**. The change in injection quantity of the pre-injection at IMEP 7 bar would lead to significantly reduced lambda after EOI. However, for this operating point, the EGR rate also had to be reduced due to soot formation and incomplete combustion, which is why the increased boost pressure (and exhaust backpressure) brings the combustion air-fuel ratio back into a comparable range. It can be deduced from this operating point that partially premixed combustion delivers the best possible results at an air-fuel ratio above 1.4 to 2.0. An EGR rate of below 30 % cannot be used with positive effects due to the early combustion phasing and rapid rise in ROHR.

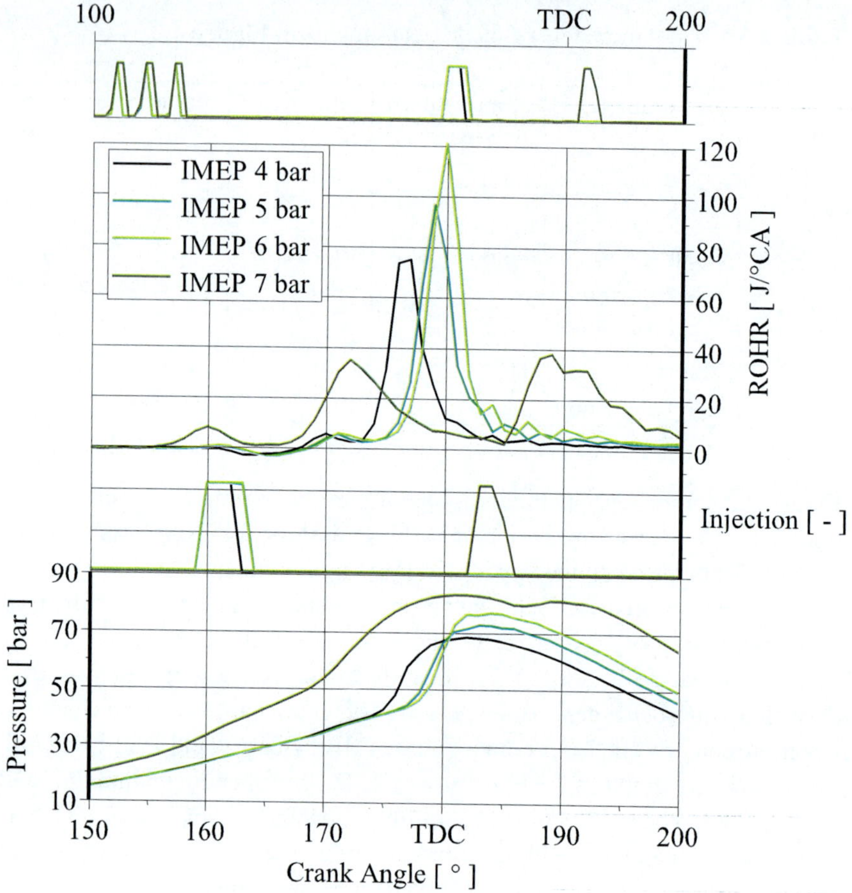

Figure 4.2: Four "Best Point" PCCI Injection Strategies at 1500 min^{-1} and IMEP of 4, 5 6 and 7 bar, Respectively.

With the same conditions for exhaust gas recirculation, intake and exhaust gas back pressure and intake temperature, a similar injection strategy for 4, 5, and 6 bar IMEP result in a different SOC of the main combustion. The author suspects a positive effect here due to lower evaporation enthalpy of the main injection, which is why radical formation leads more quickly to blue flame and thermal explosion at 170 °CAbTDC. This illustrates the effect of the small change, since here only the global air-fuel ratio is varied by the injection quantity.

In the following figures, investigations concerning PCCI are depicted. Each graph shows the average over 199 cycles, and the associated calculation of the ROHR is accomplished with TIGER from EnginOS GmbH. Also, a gas exchange analysis is calculated with intake and exhaust pressure transducers for all measurements and used for the three-stage Arrhenius described in section 5. Thus, gas exchange is not required to quantify of the PCCI combustion; The results are not displayed in the following diagrams.

4.3.1 Pre-Injection Variation

This injection sweep is mainly to analyze and understand the pre-injections on self-ignition dependency on self-ignition behavior; therefore, the EGR rate is set to zero. Other pre-injection sweeps were performed using different recirculation rates and varying parameters. The test parameters for this SOI variation are:

Table 4.4: Test Parameters of the Measurements in **Figure 4.3** and **Figure 4.4**.

Test Parameter	Abbreviation	Unit	Value
Engine Speed	n	[min^{-1}]	1500
Engine Load	IMEP	[bar]	5
Air/Fuel Ratio	λ	[-]	2.2
MFB50	MFB50	[°CAaTDC]	10
Intake Pressure	p_2	[bar]	AMB
Exhaust Pressure	p_3	[bar]	AMB
EGR Rate	EGR	[%]	0
Intake Temperature	T_2	[°C]	42

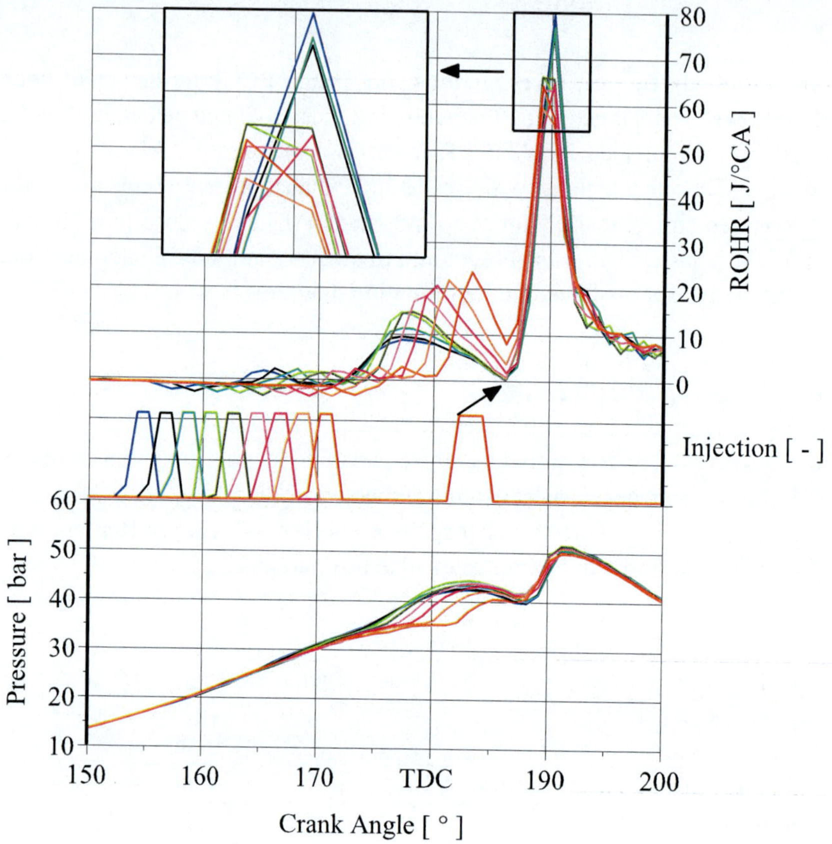

Figure 4.3: PCCI Strategy with Pre-Injection Variation at 1500 min^{-1} and IMEP of 5 bar.

Figure 4.3 shows a two-stage ignition of the pre-injected fuel. The first relevant change in ROHR starts after SOI and results in heat losses due to evaporation. This negative ROHR becomes positive at 174°CA, followed by an NTC area. This first heat release is highly dominated by the physical ID, as retarding the pre-injection affects the SOC of the LTHR (see **Figure 4.3**). The pre-injection sweep does not affect the maximum in-cylinder pressure around TDC. However, the SOI does affect the ID of the HTHR, including the maximum pressure rise and the maximum heat release per degree crank angle (as seen in the ROHR).

In conclusion, the influence on SOC of pre-injections is limited by advancing the SOI (LTHR). For the blue and black graphs in the right-hand diagram, a stagnation of the start of combustion can be observed. A further advance of the pre-injection follows in an extended ID – not in an earlier SOC. The stagnation of chemical processes primarily influences this ID. However, the conditions in the cylinder at the point of the intake valve close affect the global and local chemical processes. EGR rate, air temperature, and air pressure (and associated gas exchange) are external conditions that affect the SOC. Moreover, the internal parameters of engine speed and load, and consequently the air-fuel ratio, injection pressure, and injection strategy, influence the in-cylinder conditions such as turbulence, mixture formation for premix quality, and fuel-wall interaction. As a result, different test conditions cause complex processes that must be investigated separately.

4.3.2 EGR Rate Variation

The graphs in **Figure 4.4** illustrate an EGR variation with 0, 20, and 30 % external EGR, with the black curve illustrating the same measurement as the black curve in the plots of **Figure 4.3** (0 % EGR).

Table 4.5: Test Parameters of the Measurements in **Figure 4.4**.

Test Parameter	Abbreviation	Unit	Value
Engine Speed	n	[min^{-1}]	1500
Engine Load	IMEP	[bar]	5
Air-Fuel Ratio	λ	[-]	1.2–2.0
MFB50	MFB50	[°CAaTDC]	10
Intake Pressure	p_2	[bar]	AMB
Exhaust Pressure	p_3	[bar]	AMB
EGR Rate	EGR	[%]	0–30
Intake Temperature	T_2	[°C]	42

The test conditions listed in **Table 4.4** are also applied for these measurements, except for the EGR rate, which is shown and assigned separately in the upper left of **Figure 4.4** (shown for completeness in **Table 4.5**). For the EGR variation, the total injected fuel was held constant. Consequently, lambda (air-fuel ratio) varies from 1.5 (30 % EGR) to 2.2 (0 % EGR) for the measurements due to inert gas recirculation and decreases as exhaust gas recirculation increases.

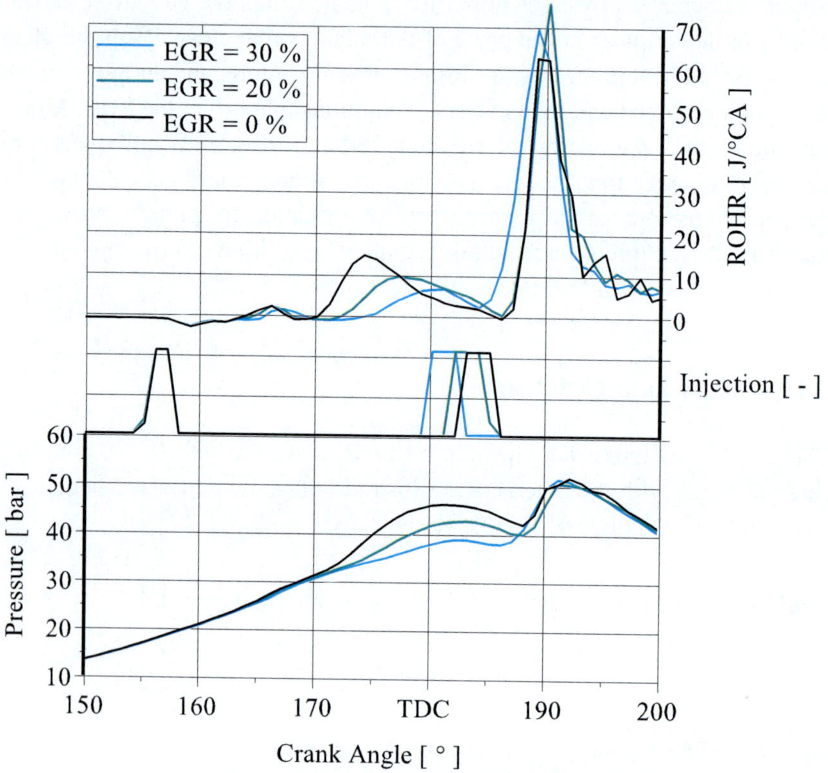

Figure 4.4: EGR Variation and Main-Injection Adjustment for Constant MFB50 with Injection Strategy of **Figure 4.3** (Black Line, 0 % EGR) at 1500 min^{-1} and IMEP 5 bar.

The ID and the amount of burned fuel before the main injection can be seen from the in-cylinder pressure diagram and in ROHR. While the ID for the

LTHR and the HTHR extends with increasing EGR rate, the cumulated rate of heat release distinguishes relevantly for premix combustion. However, the total heat release is comparable for all three measurements when the main injection is used for combustion phasing (in this case, 10 °CAaTDC MFB50 was applied). In summary, the author would like to emphasize that the deceleration of chemical reactions by exhaust gas recirculation is more dominant than the global and local decrease in the air-fuel ratio. Furthermore, the HTHR is formed in different ways. While between 20 and 30 % EGR rate, the heat release is symmetrical. For the 0 % EGR rate, the rate of heat release rises fast and decreases slow linearly. This means that a rise in the EGR rate is associated with slower (start of) burning. As burning for 0 % EGR has the same length, the second half of the ROHR (HTHR) is similar to a slow burn rate at the end of conventional diesel combustion. A rising recirculation rate is, therefore, beneficial for a symmetrical ROHR.

4.3.3 Main-Injection Variation

In addition, a sweep of the main injection can be seen in **Figure 4.5**. For each measurement, three pre-injections and one main injection were applied, with the most advanced main injection threshold being the combustion noise. Further retarding of the main injection was not relevant since the main injection of the most retarded measurement already occurs after the end of HTHR. The influence of the introduced heat by pre-injections on the main injection can be explained based on the measured pressure curves.

At the top of **Figure 4.5**, the pulses for the electrical control of the injector nozzle are shown in the relevant crank angle sector between 100 and 200 °CA. The diagram underneath is comparable to those of **Figure 4.3** (black line) and **Figure 4.4**. Due to the temporal overlap of the main injection and HTHR (pre-injection), a relevant part is highlighted on the left-hand side. This contains the four plots with the most advanced main injection, as they are in the background of all other ROHR plots and, therefore, not sufficiently visible. The LTHR is not visible in the ROHR but has a similar ID and ROHR for all measurements, except for slight variations from cycle to cycle.

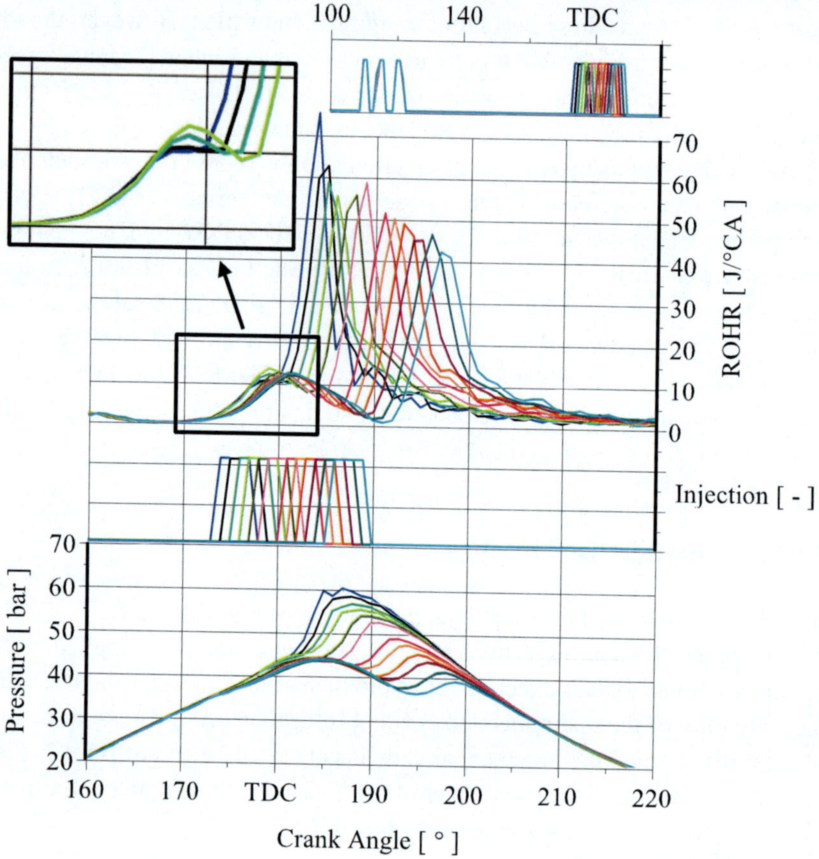

Figure 4.5: Main-Injection Variation with Three Pre-Injections and One Main at 2000 min^{-1} and Constant IMEP of 5 bar.

The onset of high-temperature combustion and, thus, the ID are not affected by the main injection. However, the ROHR is highly influenced by injection energy, evaporation, and local enrichment (see highlighted window). The blue and black plots show a clear plateau in ROHR due to the influences mentioned above, and an NTC region is not evident. However, an NTC region is clearly present in the other measurements shown in the figure. This means that the advanced main injection overcompensates the HTHR by evaporation and heat release. The effect of the main injection on the ID can be neglected in these measurements, as the ROHR and the pressure indication are evenly spaced (comparing the individual measurements) for the ROHR and pressure rise.

It follows that successive injections affect the ongoing heat release of the pre-injections. In contrast, the contribution of the pre-injections does not simultaneously impact the ID of the main injection. Hence, the multiple injection strategies interfere in many cases, and the influences become visible in offline evaluation and comparison of the parameter variation. Therefore, online evaluation is usually challenging due to the interferences of the injections.

Table 4.6: Test Parameters of the Measurements in **Figure 4.5**.

Test Parameter	Abbreviation	Unit	Value
Engine Speed	n	[min^{-1}]	2000
Engine Load	IMEP	[bar]	5
Air-Fuel Ratio	λ	[-]	1.5
MFB50	MFB50	[°CAaTDC]	4–17
Intake Pressure	p_2	[bar]	AMB
Exhaust Pressure	p_3	[bar]	1.000
EGR Rate	EGR	[%]	30
Intake Temperature	T_2	[°C]	45

The introduced variations of the pre-injection, EGR rate, and main injection represent the methodology of the investigations regarding PCCI.

4.3.4 Intake Air Temperature Variation

Another decisive factor for the ignition delay and the subsequent combustion is the question: What influence does the intake air temperature have?

To clarify this question, different intake air temperatures were set across all measuring points, and a different parameter was adjusted in each case. Overall, however, the ignition delay model in chapter 5 covers a range of T_2 between

40 and 60 °C. This also seems to be a reasonable range for series on-road application. With significantly increased EGR flow, it is impossible to set a temperature below 40 °C despite multiple cooling of the fresh intake air and the mixture of fresh air and recirculated exhaust gas. Above 60 °C, the rise in pressure of the main combustion for the operating point shown is above the limit of the engine mechanics. Nevertheless, it can be seen that the increased intake temperature causes the effective charge in the combustion chamber to drop (which is why lambda decreases at higher temperatures).

Table 4.7: Test Parameters of the Measurements in **Figure 4.6**.

Test Parameter	Abbreviation	Unit	Value
Engine Speed	n	[min^{-1}]	1500
Engine Load	IMEP	[bar]	4.8–5
Air-Fuel Ratio	λ	[-]	1.0–1.1
MFB50	MFB50	[°CAaTDC]	7–10
Intake Pressure	p_2	[bar]	AMB
Exhaust Pressure	p_3	[bar]	1.020
EGR Rate	EGR	[%]	50
Intake Temperature	T_2	[°C]	60–40

Contrary to **Figure 4.4 (Table 4.7)**, the exhaust pressure is raised by 20 mbar to achieve an EGR rate of 50 %. A slight change from 10 to 7 °CAaTDC MFB50 occurs with the higher initial energy after IVC. The parameter sweep started with 40 °C as the intake temperature and IMEP of 5.0. All parameters were fixed, and the intake temperature increased for each measurement. At 50 °C, the THC and CO emissions increased slightly, and IMEP was −0.13. Further, an increase to 55 °C results in an IMEP of −0.17, and at 60 °C, an IMEP drop of −0.2. THC and CO emissions increase, respectively. The time scales of the chemical reactions do not increase linearly with the intake temperature.

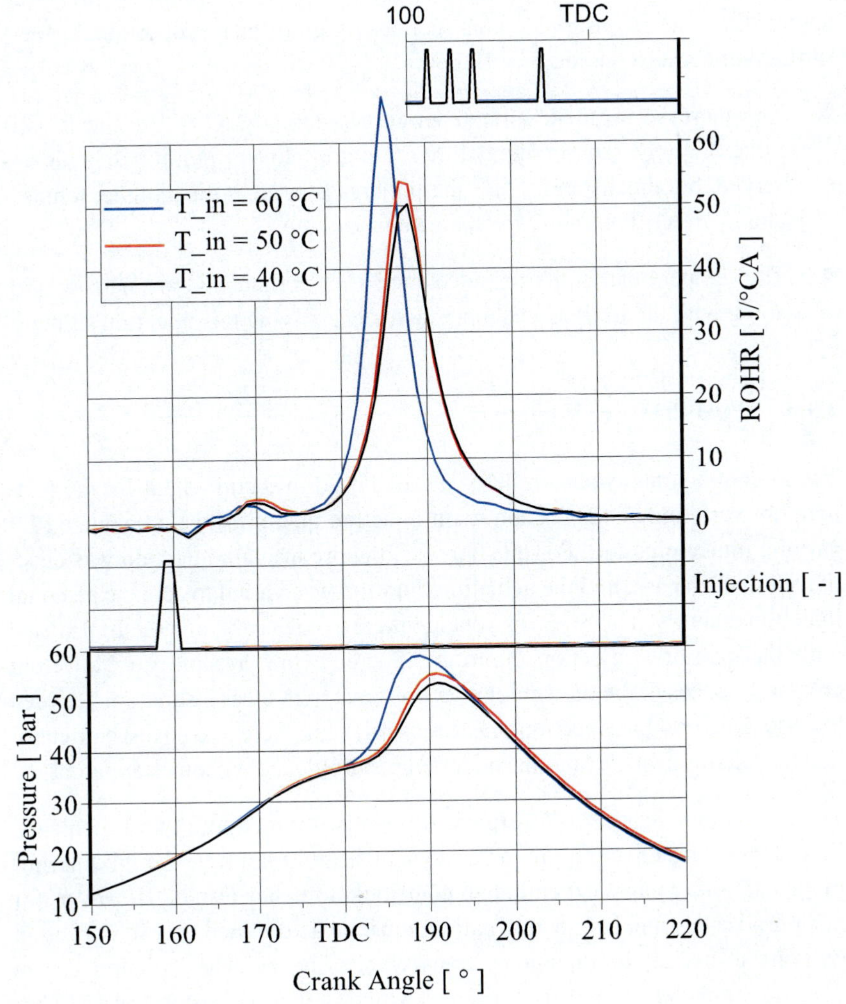

Figure 4.6: Intake Temperature Variation from 40 to 60 °C at 1500 min^{-1} and Variable IMEP. Inlet Air Density Effects the Air-Fuel Ratio.

For an increase in temperature from 40 to 50 °C, an increase in the premixed combustion rate is observed. This increase is more than proportionate for a further temperature increase of 10 degrees.

- The increase in intake air temperature is directly related to the ROHR rather than the ignition delay time. A slightly longer ignition delay is observed, but the difficulty of a sharp increase in pressure at higher temperatures affects the ability to change temperature.
- The temperature limits for adequate PCCI combustion are narrow and, along with the EGR rate, a main factor in applying this injection strategy.

4.3.5 Fuel Pressure Variation

The injection strategy shown here was also used in section 4.3.4 for the temperature variation. The pre- and main injection quantities were equalized for varying intake pressure. For this purpose, the pre or main injection was deactivated in each case, and the actuation duration was varied to achieve an equal injection quantity. The SOI was kept identical in each case, and the EOI varied with the quantity (injection duration). Thereby, the possibility of different cross-influences of the three pilot injections exists. A clear distinction between the 400 and 1000 bar injection pressure is relevant. Here, the positive effect of better fuel atomization (and introduction of turbulence) becomes apparent.

The measurements shown in **Figure 4.7** can be divided into two statements. First, increasing injection pressure can significantly improve the mixture homogenization of early injection and contribute to better results. However, for this purpose, a suitable injection pattern must be determined for each injection pressure to utilize the injections' transverse influence. The variation shown does not consider the injection timing, which is why the result for 1200 bar injection pressure is negative.

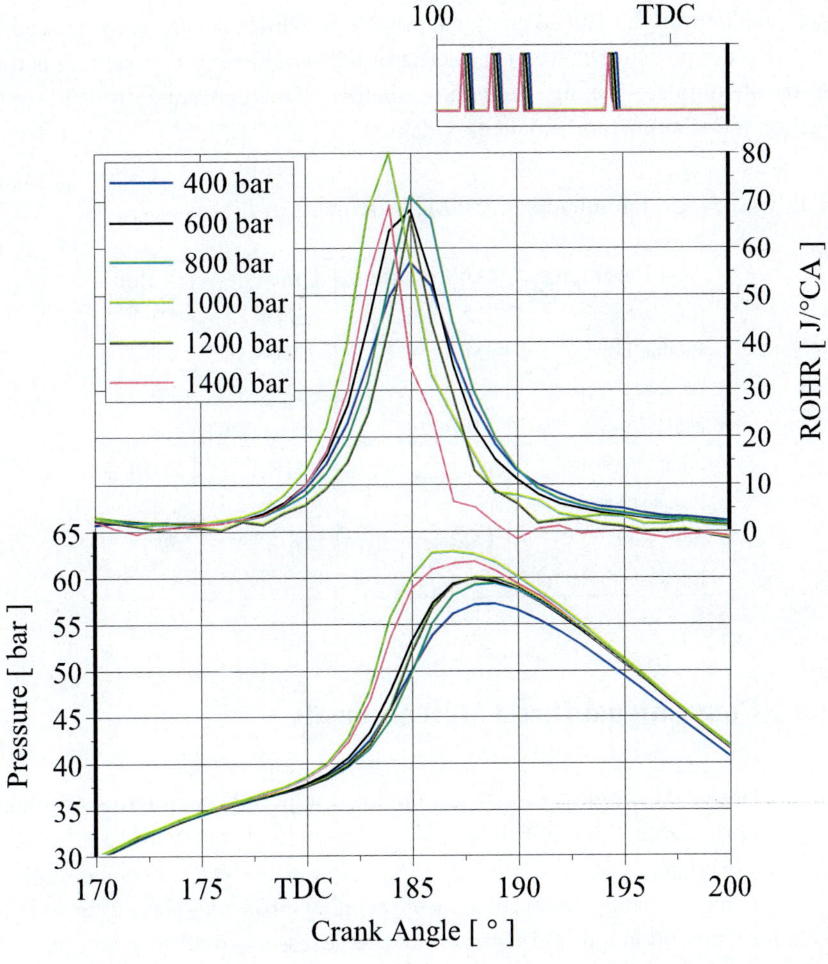

Figure 4.7: Fuel Pressure Variation from 400 to 1400 bar at 1500 min^{-1} and Variable IMEP (5 bar IMEP for 1000 bar Inj. Pressure). Equal Pre- and Main-Injection Quantity; Adjusted with Injection Duration (for Pre- and Main-Injection).

For 1200 bar, there is a significant increase in soot emissions. This is due to negative influences caused by fuel injections. The increased soot formation is noticeable through reduced chemical processes, which can certainly be seen in the cylinder pressure curve. The inertia of SOI and the conversion rate is due to the incomplete mixture preparation before SOC, consequently leading to incomplete combustion and soot oxidation.

Table 4.8: Test Parameters of the Measurements in **Figure 4.7**.

Test Parameter	Abbreviation	Unit	Value
Engine Speed	n	[min^{-1}]	1500
Engine Load	IMEP	[bar]	4.8–5
Air-Fuel Ratio	λ	[-]	1.0
MFB50	MFB50	[°CAaTDC]	5–8
Intake Pressure	p_2	[bar]	AMB
Exhaust Pressure	p_3	[bar]	1.025
EGR Rate	EGR	[%]	50
Intake Temperature	T_2	[°C]	45

4.4 Conventional Diesel Measurements

In the case of the single-cylinder test engine, the periphery in terms of intake and exhaust plenum, the EGR line, and the mechanical parts used (e.g., piston, cylinder head, and camshaft) are not comparable with the test bench compositions of the available measurements from other projects. Therefore, engine tests for conventional diesel combustion are necessary to allow a comparison of engine performance, emissions, and other relevant parameters with the PCCI configuration. For the interests of comparability, the “best point” focusing on engine exhaust emissions and fuel consumption – defined by the indicated (ISFC) or brake-specific fuel consumption (BSFC) – was determined by a full-factorial variation of the key parameters at the test bench.

As a specification, one pilot and one main injection were applied. In addition, further measurements with PCCI in terms of low emissions and BSFC were investigated to compare these measurements of conventional diesel injection with the PCCI strategy.

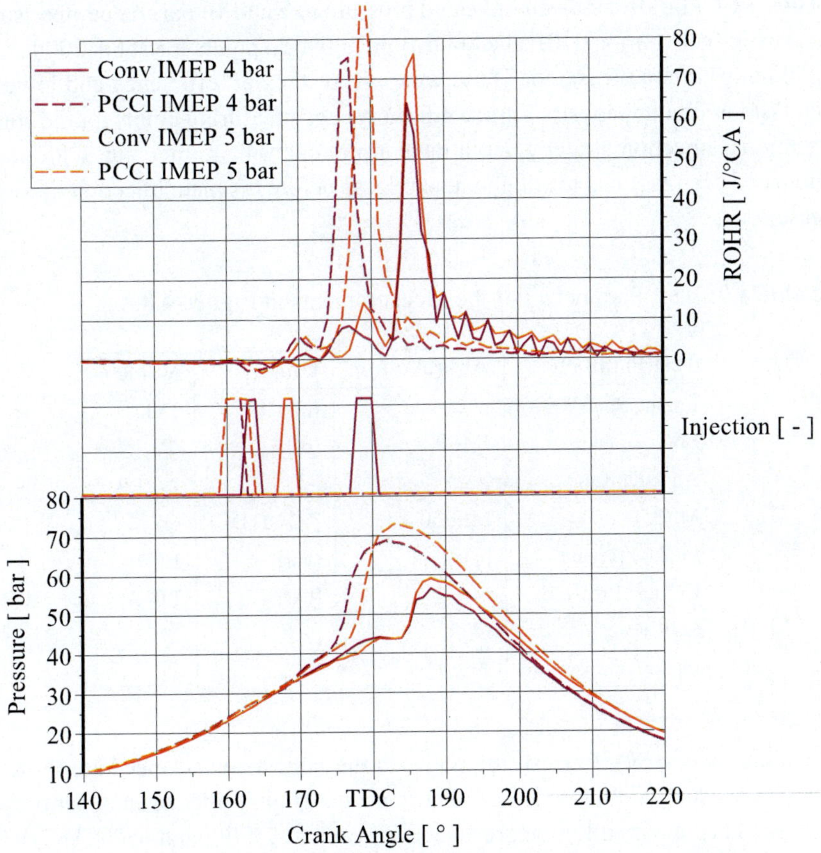

Figure 4.8: Two Conventional Diesel Injection Strategies at 1500 min^{-1}. One Pilot and One Main-Injection with Pilot Injection Variation. Additional Dashed Lines: PCCI with IMEP of 4 and 5 bar for Comparison.

A conventional diesel injection strategy is shown in **Figure 4.8**. The simplification of the state-of-the-art injection as used in actual applications is done as described above. A load sweep concerning IMEP was performed for the conventional tests. The indicated mean pressure delta is 1 bar, and the range spans from 4 to 7 bar and 4 to 10 bar for 1500 and 2000 min^{-1}, respectively. Thus, PCCI limits the mean indicated pressure to 7 and 10 bar. As comparison points for conventional diesel combustion, only best points for 1500, 2000 and 2500 min^{-1} were measured. These were set for the best efficiency and lowest emissions simultaneously. **Figure 4.8** shows two measuring points for the conventional injection strategy. Both have a pre and a main injection, with pronounced pre-reactions leading to heat release before the main injection in each case.

Table 4.9: Test Parameters of the Measurements in **Figure 4.8**.

Test Parameter	Abbreviation	Unit	Value
Engine Speed	n	[min^{-1}]	1500
Engine Load	IMEP	[bar]	4/5
Air/Fuel Ratio	λ	[-]	2.6/1.9
MFB50	MFB50	[°CAaTDC]	6/7
Intake Pressure	p_2	[bar]	1.070
Exhaust Pressure	p_3	[bar]	1.042
EGR Rate	EGR	[%]	29
Intake Temperature	T_2	[°C]	42

In addition, two PCCI measuring points at the same load and speed are shown in the diagram with dashed lines. These have a slightly increased boost pressure of 1110 mbar and three pre-injections each at 75, 70, and 65 °CAbTDC. The comparison of emissions and noise can be read from the diagrams in section 4.5. However, a direct comparison of the in-cylinder pressure and ROHR shows a significant difference. For the main injection of PCCI, a very early point must be selected to suppress soot formation. This leads to early ROHR before and at TDC.

On the other hand, the main injection of the conventional strategy is not triggered until TDC, which is why the pressure-forming component is formed after TDC due to the ignition delay. This leads to an MFB50 of 6 and 7 °CAaTDC and, therefore, almost optimal. Of course, this can also be seen in the maximum pressure of the combustion since this remains significantly lower (below 60 bar) due to the later combustion. However, this mechanical stress is not damaging the conventional diesel engine since the engine is designed for approximately 175 bar peak pressure. Nevertheless, early combustion is also not advantageous in the pV diagram, which is why unnecessary combustion noise is introduced into the structure.

4.5 Evaluation, Discussion and Comparison of PCCI and Conventional Diesel Injection Strategies Concerning Combustion and Engine-Out Emissions

The direct comparison in this chapter contains only measurements from the single-cylinder test bench at the IFS of the University of Stuttgart. For this purpose, the test bench and the engine were neither disassembled nor modified between the individual measurements. The results are, therefore, directly comparable. An absolute comparison with other publications in which a Mercedes Benz OM642 engine was used can explicitly not be recommended since the mechanical parts used in this configuration are not assembled in the same configuration by Mercedes Benz. Full factorial changes were made to several parameters on the test bench for the measurements shown here. Good results were achieved, but these cannot be regarded as the absolute best point for either the conventional combustion process or PCCI. Furthermore, only steady-state points are shown, so a direct comparison with transient measurements is impossible.

The following diagrams compare the conventional and PCCI combustion strategies regarding THC, CO, NO_X, FSN (soot), EGR rate, and fuel consumption (BSFC). The selected engine speed in **Figure 4.9** and **Figure 4.10** is 1500 and 2000 min^{-1}, and the engine load is IMEP 5 and 6 bar, respectively. The normalization is accomplished for each measurement and each point shown. For example, the CO emission of PCCI at 1500 min^{-1} and IMEP 6 bar is 3579 ppm; for the conventional strategy and the same engine speed and load, the CO emission is 645 ppm. Therefore, the most valuable value – in this case, the 3579 ppm of the PCCI measurement – is defined as “1”, and the least valuable value – in this case, the 645 ppm of the conventional measurement – is set in relation to the 3579 ppm. Therefore, PCCI is "1" and the relative CO emission for conventional is "0.18" (see **Figure 4.9**, right).

Additionally, the EGR rate illustrates the required quantity of EGR to achieve low NO_X and soot emissions and provide sufficient ID for the premix conditions and combustion phasing. For efficiency, BSFC is used to compare the strategies. Despite the single-cylinder engine, where friction losses are not comparable to those of the series engine, the BSFC is most predictable due to external pressure regulation (intake and exhaust pressure) and the associated variations in the calculation of the indicated mean effective pressure. Hence, a percentage comparison of BSFC with the same engine is more expressive and accurately represents the conditions.

In conclusion, across <u>all</u> measurements, PCCI has its benefits in NO_X and filter smoke number (soot). However, the conventional diesel strategy has lower fuel consumption which translates directly into lower CO and THC emissions. This fuel consumption advantage reduces to about 5 % in some cases – and, as mentioned earlier, the parts in the cylinder were designed to work in a conventional diesel application. Piston (combustion chamber) geometry and an optimized injector could result in less wall impingement and comparable fuel consumption for PCCI. It is known from the literature that improved fuel consumption and thus lower emissions (mainly CO and THC) could be achieved with the swirl flap in both PCCI and conventional diesel combustion, since the swirl (along the cylinder axis) and quenching results in increased turbulence and, thus, better mixing of fuel and air. However, this was not investigated here as there were sufficient other variation parameters available that were designated to be more important for evaluating the ID of PCCI.

The relative comparison of some best points of partially premixed and conventional diesel combustion can be seen in **Figure 4.9** and **Figure 4.10**. Furthermore, an absolute comparison of THC, CO, NOx, soot, and fuel consumption is shown in **Figure 4.11**. In addition, the external EGR rate is shown, with significantly higher values for PCCI at 1500 and 2000 min^{-1} for all engine loads. On the one hand, this leads to worse combustion; on the other hand, a long ignition delay is achieved to avoid advancing combustion phasing too early. For the load sweep at 1500 min^{-1}, partially premixed combustion could only be applied up to IMEP 7 bar. Above this, combustion becomes too unstable, resulting in significantly increased emissions.

The quality of the characteristic values will be discussed again. The measuring points were all reached in the steady-state operating condition. However, no direct comparison with transient engine operation is possible. Auto-ignition and combustion exclusively depend on the composition in the combustion chamber, with some operating points being run at the limit of possibility. A small change (e.g., in exhaust backpressure) causes combustion to collapse, as too much or too little exhaust gas is recirculated internally and externally. Rapid switching from conventional to partially premixed combustion is thus only possible to a limited extent since combustion stability is not necessarily given due to many parameters. For this, a smooth transition would have to be applied, which is why a switchover is expected to have a smaller advantage in terms of nitrogen oxide and particulate emissions.

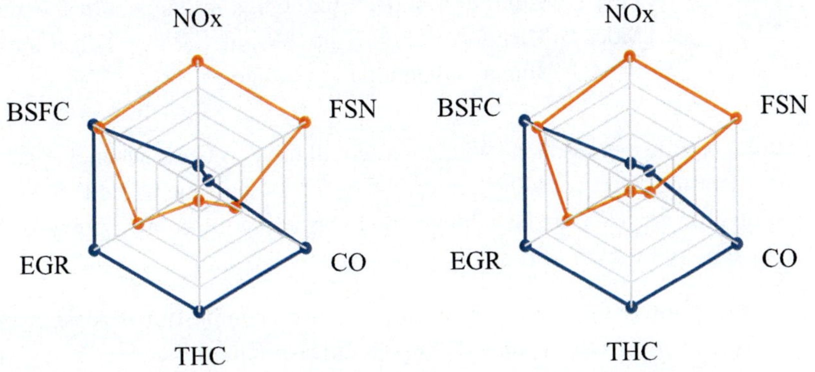

Figure 4.9: Direct Comparison of Conventional and Alternative Combustion Processes at 1500 min^{-1} in Terms of EGR Rate, Specific Fuel Consumption and Engine-Out Emissions. On the Left Side, IMEP 5 bar, on the Right Side, IMEP 6 bar. PCCI = Blue, Conventional = Orange.

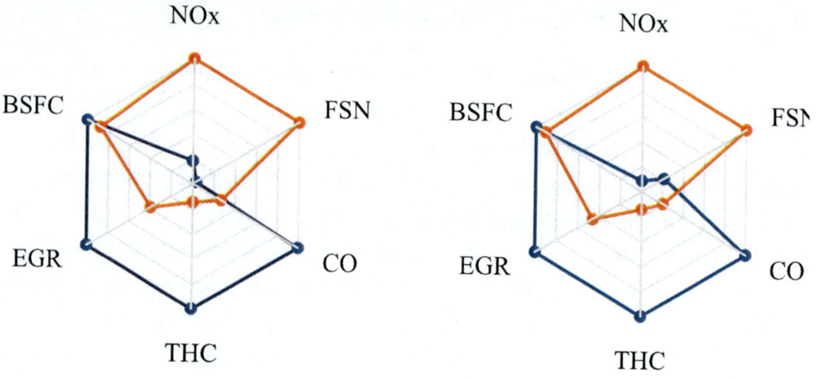

Figure 4.10: Direct Comparison of Conventional and Alternative Combustion Processes at 2000 min^{-1} in Terms of EGR Rate, Specific Fuel Consumption and Engine-Out Emissions. On the Left Side, IMEP 5 bar, on the Right Side, IMEP 6 bar. PCCI = Blue, Conventional = Orange.

For real engine operation, a switchover would have to occur whenever an engine operating point could be held constant for longer time in the PCCI combustion process. This can be achieved by hybridizing the entire system to shift the load point in favor of partially premixed combustion.

The description of the engine noises introduced by combustion is possible both directly and indirectly. Either the signal (measurable, objective) or sensing (psychoacoustic, subjective) magnitudes can be applied for the determination. The results of a combustion analysis via combustion parameters such as maximum peak pressure, maximum pressure gradient, and mean pressure are decisive for evaluating the various combustion processes. The combustion noise serves as a further evaluation criterion for combustion and combines and assesses the characteristic values. The measured cylinder pressure, crank angle, and engine speed determine this direct combustion noise. For this purpose, a frequency transformation is first performed using the Fast Fourier Transform (FFT). The result of the cylinder pressure excitation spectrum must then be transferred. The transfer with the aid of the structure-transfer measure, which is already A-weighted, leads to direct combustion noise. This structure transfer measure has been defined by measurements of different engines as a standard structure transfer measure and is available in the literature.

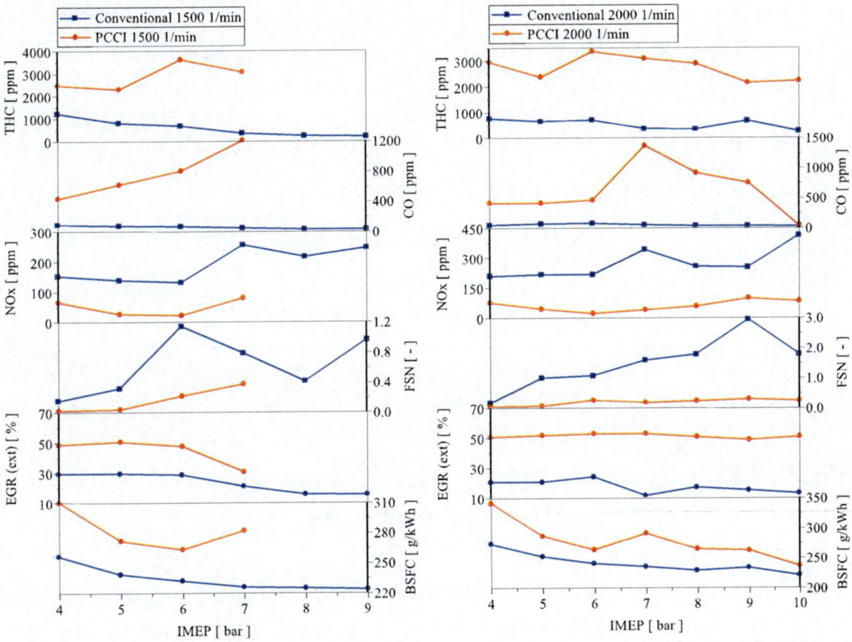

Figure 4.11: Absolute Comparison of PCCI and Conventional Diesel at 1500 and 2000 min^{-1}.

Figure 4.12 shows the cylinder pressure curve and the associated cylinder pressure spectrum (without structure transfer coefficient). The main frequency range is shown on the figure's right, where n is given in revolutions per second. The low-frequency range is thus significantly dependent on the maximum cylinder pressure. Increasingly, the pressure gradient is decisive up to 40n (1500 min^{-1} = 1000 Hz), and the pressure gradient is decisive above that.

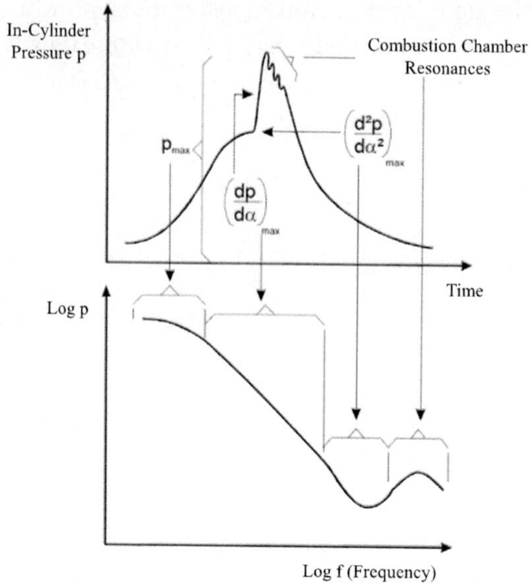

p_{max}	$f_{limit} \approx 10 \cdot n$
$\left(\frac{dp}{d\alpha}\right)_{max}$	$f_{limit} \approx 40 \cdot n$
$\left(\frac{d^2p}{d\alpha^2}\right)_{max}$	$f \geq 40 \cdot n$
Combustion Chamber Resonances	$f \approx 4 - 7$ kHz

n = Engine Speed [min^{-1}]

Figure 4.12: Exemplary Cylinder Pressure Curve and Cylinder Pressure Spectrum (Translated from [114]).

The following three diagrams compare partially premixed and conventional diesel combustion, respectively. The first diagram shows measurement points at 1500 min^{-1} and a load of IMEP 4–7 bar from left to right. The second and third diagrams contain measuring points at 2000 min^{-1} and IMEP 4–10 bar in ascending order. In the top section of the diagram, the combustion noise (A-weighting with standard structure transmission dimension) for PCCI is black. Conventional diesel combustion is shown in gray, with the combustion noise plotted against the logarithmically plotted frequency. The pressure curve is shown with a solid line in the middle and bottom graphs, and the pressure rise gradient is shown in dotted lines. The middle graph contains measurements of conventional combustion, and the bottom graph contains measurements of partially premixed combustion. Below each diagram, a table is shown. This contains the combustion characteristics of the measuring points shown, whereby PCCI and conventional combustion are not one above the other but side by side. In each case, the maximum rise in pressure of the mean value (199 cycles), the maximum rise in pressure of one (maximum) operating cycle, the maximum cylinder pressure and the position in degrees of crank angle are plotted. This shows that conventional diesel combustion is also assumed to be valid for these measurement points (best points) with a high-pressure rise gradient. For these measurements, the focus was on a boundary analysis. In some cases, the maximum cylinder pressure and the pressure rise gradient were measured to be twice the value for PCCI.

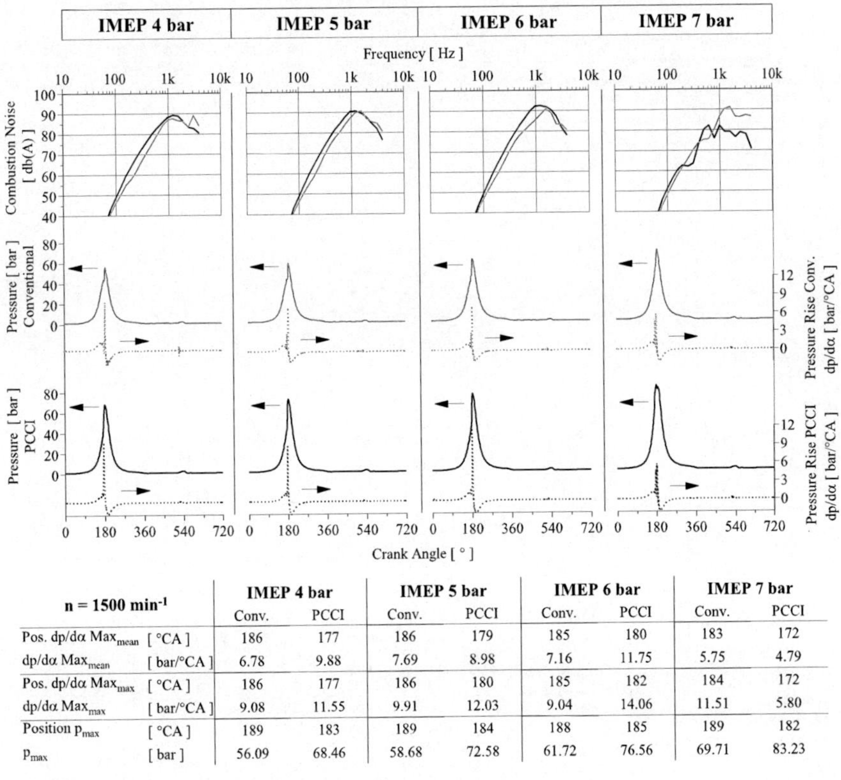

$n = 1500\ min^{-1}$		IMEP 4 bar Conv.	IMEP 4 bar PCCI	IMEP 5 bar Conv.	IMEP 5 bar PCCI	IMEP 6 bar Conv.	IMEP 6 bar PCCI	IMEP 7 bar Conv.	IMEP 7 bar PCCI
Pos. dp/dα Max_{mean}	[°CA]	186	177	186	179	185	180	183	172
dp/dα Max_{mean}	[bar/°CA]	6.78	9.88	7.69	8.98	7.16	11.75	5.75	4.79
Pos. dp/dα Max_{max}	[°CA]	186	177	186	180	185	182	184	172
dp/dα Max_{max}	[bar/°CA]	9.08	11.55	9.91	12.03	9.04	14.06	11.51	5.80
Position p_{max}	[°CA]	189	183	189	184	188	185	189	182
p_{max}	[bar]	56.09	68.46	58.68	72.58	61.72	76.56	69.71	83.23

Figure 4.13: Comparison of Combustion Noise, Pressure Rise (Dotted) and In-Cylinder Pressure of PCCI (Black) and Conventional (Gray) Combustion at 1500 min^{-1} for 4, 5, 6 and 7 bar IMEP (From Left to Right).

The third-octave frequency range can be calculated by the sampling frequency of the indicated combustion pressure signal only up to 5000 Hz (1500 min^{-1}) and 6000 Hz (2000 min^{-1}). The Nyquist-Shannon sampling theorem states that equidistant sampling values must correspond to twice the frequency in order to reconstruct the true values. This results in this maximum one-third octave band for the different speeds.

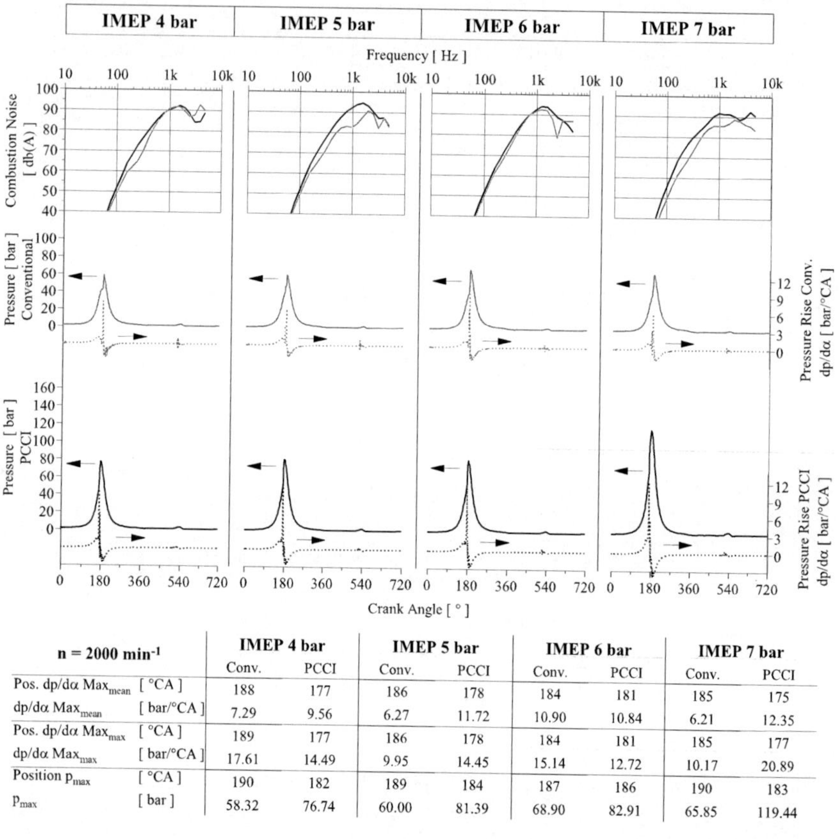

n = 2000 min^{-1}		IMEP 4 bar		IMEP 5 bar		IMEP 6 bar		IMEP 7 bar	
		Conv.	PCCI	Conv.	PCCI	Conv.	PCCI	Conv.	PCCI
Pos. dp/dα Max_{mean}	[°CA]	188	177	186	178	184	181	185	175
dp/dα Max_{mean}	[bar/°CA]	7.29	9.56	6.27	11.72	10.90	10.84	6.21	12.35
Pos. dp/dα Max_{max}	[°CA]	189	177	186	178	184	181	185	177
dp/dα Max_{max}	[bar/°CA]	17.61	14.49	9.95	14.45	15.14	12.72	10.17	20.89
Position p_{max}	[°CA]	190	182	189	184	187	186	190	183
p_{max}	[bar]	58.32	76.74	60.00	81.39	68.90	82.91	65.85	119.44

Figure 4.14: Comparison of Combustion Noise, Pressure Rise (Dotted) and In-Cylinder Pressure of PCCI (Black) and Conventional (Gray) Combustion at 2000 min^{-1} for 4, 5, 6 and 7 bar IMEP (From Left to Right).

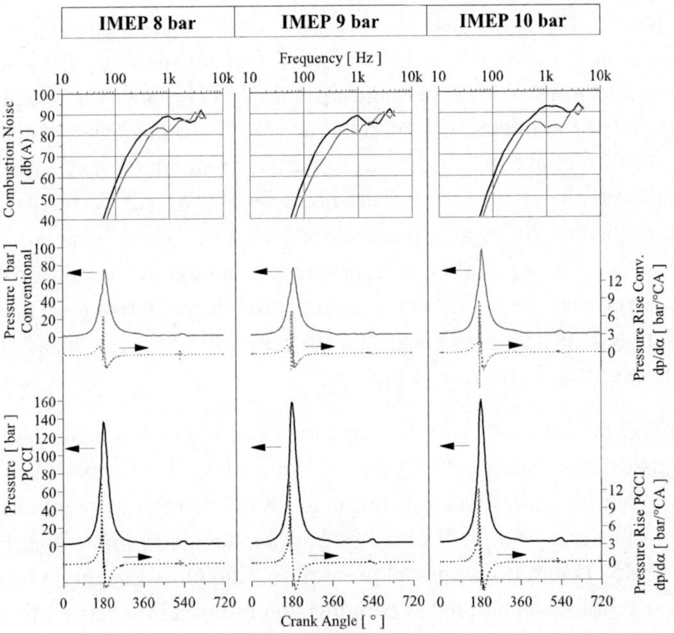

n = 2000 min^{-1}		IMEP 8 bar		IMEP 9 bar		IMEP 10 bar	
		Conv.	PCCI	Conv.	PCCI	Conv.	PCCI
Pos. dp/dα Max$_{mean}$	[°CA]	180	174	185	172	178	173
dp/dα Max$_{mean}$	[bar/°CA]	6.49	13.64	7.38	13.50	8.69	12.22
Pos. dp/dα Max$_{max}$	[°CA]	185	176	187	172	185	176
dp/dα Max$_{max}$	[bar/°CA]	12.24	19.29	13.80	17.86	12.80	26.17
Position p$_{max}$	[°CA]	191	183	192	183	190	184
p$_{max}$	[bar]	75.05	136.02	75.96	156.89	95.07	158.82

Figure 4.15: Comparison of Combustion Noise, Pressure Rise (Dotted) and In-Cylinder Pressure of PCCI (Black) and Conventional (Gray) Combustion at 2000 min^{-1} for 8, 9 and 10 bar IMEP (From Left to Right).

The combustion noise and maximum pressure rise (peak) for PCCI combustion at IMEP 6 bar is higher than at IMEP 7 bar. A slight change of parameters (auch as EGR rate and injection strategy) results in increased fuel efficiency, NO_X, CO and even FSN (compare **Figure 4.11**). The noise level in the lower frequency range (below 800 Hz) for conventional combustion is significantly reduced. Above all, the operating point of 6 bar IMEP has clear advantages here. In general, it can be observed that both the combustion noise, which occurs at frequencies well below 1000 Hz and is determined by the maximum pressure in the cylinder, and the range around 1000 Hz (pressure rise gradient) are improved for conventional combustion. On the other hand, the pressure gradients of PCCI have disadvantages in the maximum noise level, which then also determines the sum level and the noise level. The subjective perception of psychoacoustics has many parameters based on which evaluation takes place. However, the general perception predicts a negative result for the partially premixed diesel combustion in the essential frequency range.The general perception predicts a negative result for the partially premixed diesel combustion in the essential frequency range.

Furthermore, the possibility of deposit formation in the combustion chamber and the gas exchange components exists. Contrary to the theory of heavy cylinder wall wetting with diesel, this originates less from the introduction of fuel into the combustion chamber but rather from inadequate combustion. Particularly in low-load operations with a low compression ratio, this might be a problem. The used fuel and a reduced compression ratio mainly cause this deposit formation. In this respect, the fuels used in high-speed diesel engines in road traffic are partially excluded from deposit formation due to low-quality fuel. PCCI offers the possibility of almost soot-free combustion. This directly reduces deposits on the cylinder head, piston, cylinder liner, injector, and the entire air system (due to exhaust gas recirculation and backflow through negative valve overlap). Increased wear of, e.g., valve seat rings, valve guides, piston rings, and hence oil contamination can thus be reduced. Other reduction methods include increasing the compression ratio, the pressure level, and load (e.g., cylinder deactivation) or raising the temperatures of intake air, engine oil, and cooling water. According to Eilts, the water temperature has the least positive effect. [115] PCCI has a positive effect against the formation of deposits in the combustion chamber, as the combustion has significantly reduced pollutant formation. However, this advantage would have to be verified by keeping the same load point constant over a longer period of operation and then carrying out an objective inspection. The author can also imagine the negative effects caused by the wall application of diesel. Whereby the high combustion pressure can have an opposite, positive effect. Since the engine and the periphery were used exclusively in mixed operation, an assessment of the deposit formation is impossible or cannot be attributed to the combustion process.

5 Empirical Based Model to Depict Ignition Delays for PCCI

The detection of the SOC was determined using the tangent function from the previous project, which Beck published in 2012 [62]. However, instead of the burn rate, the ROHR was therefore used. Furthermore, it was established that for the detection of the LTHR, the zero crossing of the ROHR provides more accurate results. Since, theoretically, no zero crossing has to occur with the calculated burn rate (see **Figure 2.4**), this was of course not defined in the previous project. Thus, it can be summarized that the ignition start of the low-temperature combustion is defined as the zero-crossing of the calculated ROHR curve. The ignition delay of the high temperature combustion, however, is defined by the presented ignition integral, which will be described here again with additional information as they are relevant with the main injection:

- Determine the maximum of the 1st derivative (after the NTC area and before the start of the main injection).

  → Crank Angle Position [A].

- Determine the minimum before this maximum of the 1st derivative.

  → Crank Angle Position [B].

- Value of ROHR at crank angle position [A] → [C].

  Value of ROHR at crank angle position [B] → [D].

- Deviation of [C] and [D] → [E] at 25 % of the deviation.

- Zero Crossing of straight-line intersecting [E] and [C] is defined as SOC.

Furthermore, the ignition delay of diesel engines and the calculation of the associated heat release are well discussed in the literature. [116–118]

M. S. Wahl, *Emission Reduction with an Alternative Diesel Combustion Process*, Wissenschaftliche Reihe Fahrzeugtechnik Universität Stuttgart,
https://doi.org/10.1007/978-3-658-42094-9_5

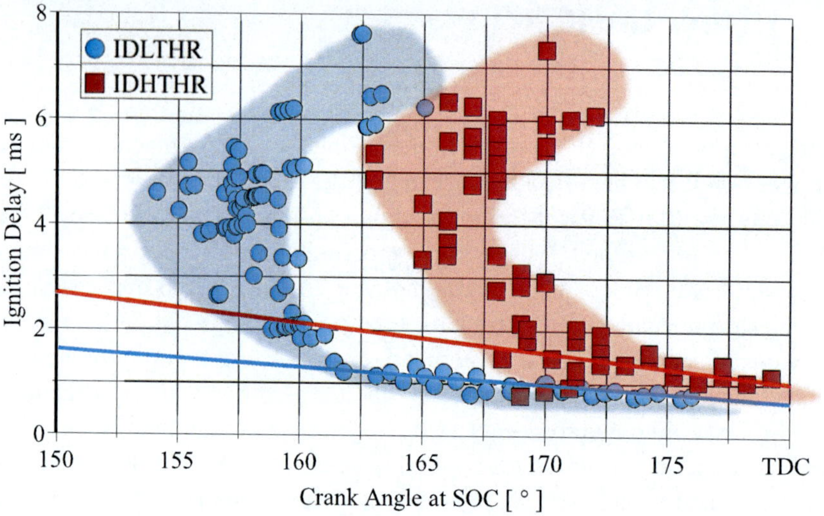

Figure 5.1: Ignition Delay of the Low and High Temperature Heat Release Starting from Pre-Injections SOI in Milliseconds.

As can be seen, the SOC depends on different parameters for both LTHR and HTHR. The influence of various parameters on chemical reactions is shown in **Figure 5.1**. Advancing the pre-injection affects the SOC (and maximum ROHR) such that the ID shows a linear dependence with increasing ignition delay for an earlier SOC. The inserted blue and red lines in **Figure 5.1** illustrates this dependency. At about two milliseconds of ignition delay, the effect of injection on SOC decreases. From two to five milliseconds ID, SOC is influenced by parameters such as EGR rate, injection pressure, in-cylinder lambda, intake air temperature, and the p_2/p_3 ratio. At 4.5 milliseconds LTHR ID, a wide spread of SOC is visible. The first ignitions start at about 154 °CAbTDC and latest at 160 °CA. This deviation of 6 °CA due to dependence on chemical reactions is a wide range, and cross-influences of the parameters mentioned above are visible. An even longer ignition delay is caused by earlier injection or a combination of lower intake temperature, higher EGR rate, lower air-fuel ratio after pre-injection, and lower intake pressure. These two ways of extending the ignition delay result in the C-curve (area in blue and red in **Figure 5.1**) as of the ignition delay for both low and high-temperature ignition.

As described in section 1.5, a literature-based three-stage Arrhenius approach was optimized to calculate the ID of the low and high-temperature combustion for pre-injections. The calculation of the ignition integral in Eq. 1.3 using a fitted Arrhenius approach and the three stages of equation Eq. 1.4 leads to the results shown in **Figure 5.2** and **Figure 5.3**.

Firstly, hyperparameter optimization was performed on a training data set to find a suitable parameter set to calculate the ID of the selected data. Then, this training set was randomly selected from the measured data set to achieve maximum variability for the model's training data. As a result, the following equation was established:

$$\tau_{ID,i} = A_i \cdot \left[\frac{p_j}{p_{ref}}\right]^{ß_i} \cdot \exp\left[\frac{T_{A,i}}{T_j}\right] \cdot \lambda^{y_i} \cdot \exp\left[\frac{EGR \cdot c_i}{T_j}\right] \qquad \text{Eq. 5.1}$$

The author's pre-published papers set the equation to the following:

$$\tau_{ID,i} = A_i \cdot \left[\frac{p_j}{p_{ref}}\right]^{ß_i} \cdot T_j^{\,b_i} \cdot \exp\left[\frac{T_{A,i}}{T_j}\right] \cdot \lambda^{y_i} \cdot \exp\left[\frac{EGR \cdot c_i}{T_j}\right] \qquad \text{Eq. 5.2}$$

Optimization of the parameter finding algorithm showed that the advantages of the calculation speed are more relevant than an additional pre-exponential factor. The measurement range of the intake temperature from 26 °C to 60 °C could be covered by equation Eq. 5.1 without the mass mean temperature dependency of equation Eq. 5.2.

A, *ß*, T_A, *y*, and *c* are parameters dependent on *i*. The reference pressure p_{ref} is always 1 bar and, so far, not operation point dependent. On the other hand, the in-cylinder pressure *p* and the mass mean temperature *T* are crank angle dependent (*j*). The air-fuel ratio is calculated at the end of each injection, and the EGR rate is calculated based on the CO_2 measurement in the intake and exhaust plenum – see equation Eq. 3.1. Eq. 5.1 in Eq. 1.3 and Eq. 1.4 leads to 18 parameters that must be adjusted to fulfill all ID calculations for different injection strategies for LTHR and HTHR ID.

After that, the calculation of these 18 parameters was performed with the available measurements and evaluations. Therefore, all ignition integrals were calculated with a single set of parameters. Any deviation of the measured self-ignition crank angle from the calculated one was stored and summed up. If the deviation is greater than 5 °CA, 8 °CA, or 12 °CA, different factors were used for weighting and balancing before summing up. This is useful to find a parameter set where most crank angle deviations are relatively small and fewer outliners exist. The parameter set with the smallest deviation with weighted results (summed) for all measurements has the best prediction. If the ignition integral of the Eq. 1.4 does not reach "1" before the next injection starts, the parameter set is either invalid, or ignition (of LTHR or HTHR) would occur after SOC. If a parameter set is invalid for at least one measurement, it is considered invalid for all measurements.

Table 5.1: Three-Arrhenius Parameter Setting for Eq. 5.1. Results in **Figure 5.2**, **Figure 5.3**, and **Figure 5.4**.

i	A	ß	T_A	y_1	c_i
1	1.714	−0.2857	2285.71	0	0.4285
2	1	−0.1	1000	2	−2.142
3	1e−7	2	12500	0	0

Figure 5.2 and **Figure 5.3** show the crank angle position at which each operating point's calculation and measured ignition occurs. When the calculation underestimates the ID, the dot is underneath the angle bisector. If the calculation overestimates the ID, the dot is above the angle bisector. For a single dot in the same sector, the color is dark (blue or red), and the dot is small. Larger dots will be lighter, indicating more than one calculated/measured result in that sector. Overall, the LTHR ignition delay estimate appears to be very accurate, as most parameter sweeps are included. The coefficient of determination is 0.92 and is sufficiently accurate for prediction.

For the prediction of the HTHR ID, the coefficient of determination is almost 0.8. The cross-influences to the in-cylinder conditions are further expanded by the introduced heat from the low-temperature combustion.

Compared to **Figure 5.2**, there are fewer dots in **Figure 5.3**. In some measurements, the main injection SOI thus starts before the HTHR of pre-injected fuel occurs. This results in an unpredictable ID. While an ID of the LTHR is available, there is no predictable HTHR ID. The achieved accuracy may be improved with an advanced algorithm such as Rprop+ [119] or by splitting the three-stage Arrhenius to adapt a second parameter set for selected conditions. However, the accuracy seems to be sufficient for the single parameter set.

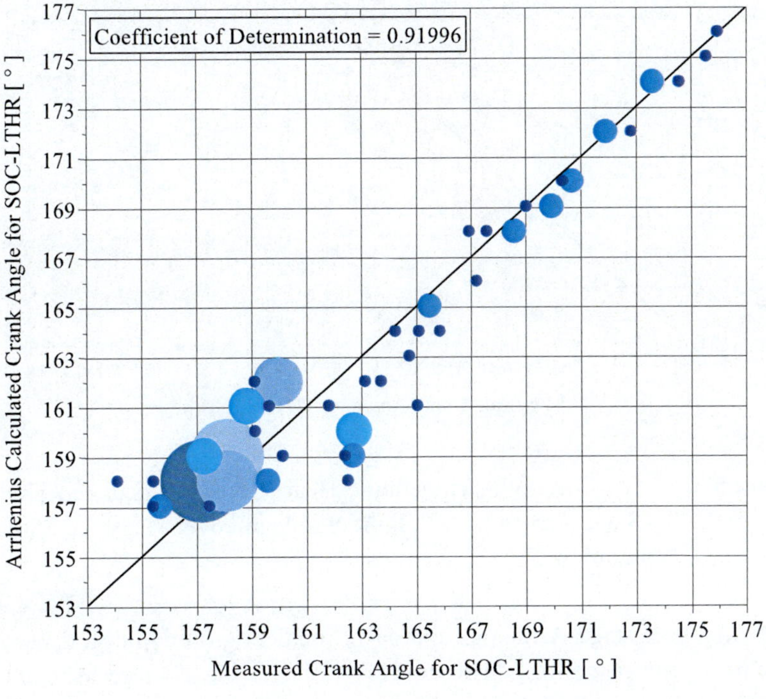

Figure 5.2: Start of Low-Temperature Heat Release: Calculated Three-Stage Arrhenius with Ignition Integral over Measured Start of Ignition [2].

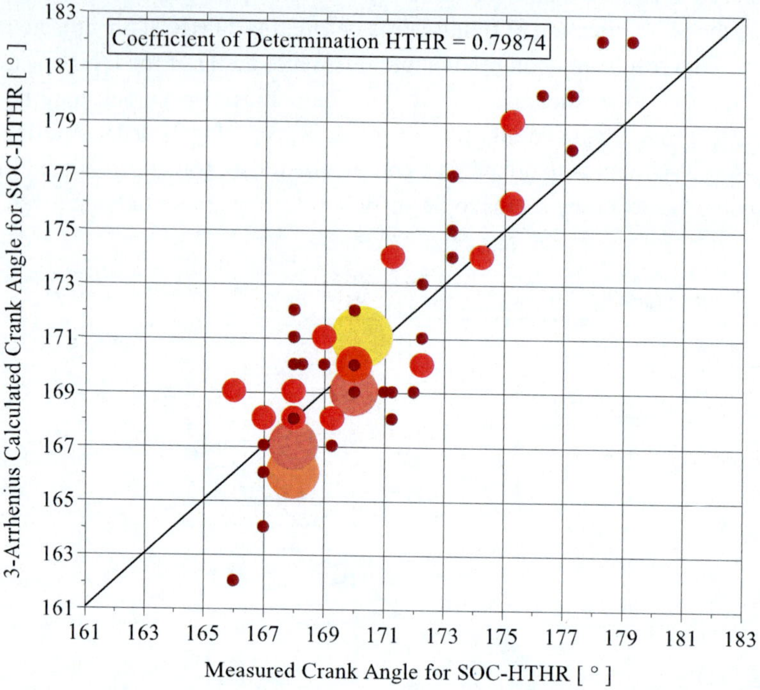

Figure 5.3: Start of High-Temperature Heat Release: Calculated Three-Stage Arrhenius with Ignition Integral over Measured Start of Ignition [2].

With this three-stage Arrhenius approach, predicting the ignition delay in a wide parameter range for LTHR and HTHR is feasible. To forecast combustion processes for different types of engines and fuel, Ivan Ivanovitch Wiebe (Vibe) published an analytic function (equation 57 in [120]) to characterize the shape of burnt fraction or burn rate:

$$w_\varphi = \frac{dx}{d\varphi} = \frac{6.908 \cdot (m+1)}{\varphi_d} \cdot \left(\frac{\varphi}{\varphi_d}\right)^m \cdot e^{-6.908 \cdot (\frac{\varphi}{\varphi_d})^{m+1}} \quad \text{Eq. 5.3}$$

The parameterization of the function by the so-called form factor m and the combustion duration φ_d leads to the well-known Wiebe burn characteristics.

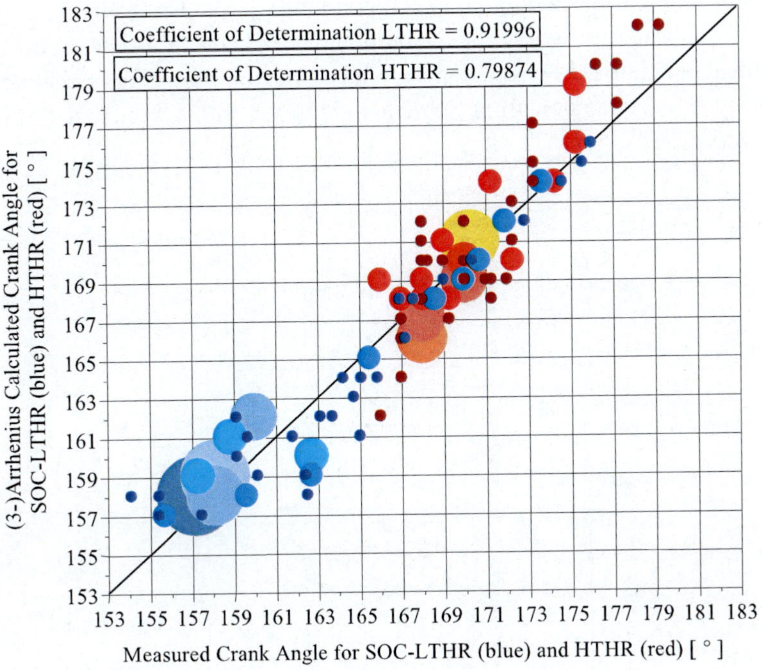

Figure 5.4: Start of Low and High-Temperature Heat Release: Calculated Three-Stage Arrhenius with Ignition Integral over Measured Start of Ignition.

For conventional diesel engines, *m* has been described as 0 to 0.7, and for gasoline engines, 3 to 4 was common [120]. Where a smaller form factor leads to a rapid increase in combustion rate at start of burning and a weakening at the end of combustion – where on the other hand a form factor above 1.5 shows a weakening of SOC and a shift of the combustion rate peak to the second half of combustion. For *m* equal to 5.0 and above, the main conversion and, thus, the peak of the burn rate is in the third quarter of the total combustion duration. The use of Eq. 5.3 to describe the heat release (not burn rate or burn fraction) requires a variable set of parameters. As in **Figure 4.4**, the influence of EGR affects the parameter *m* in the Wiebe equation. However, it is not only the EGR rate that characterizes the rate and duration of combustion; other parameters, such as dilution (depending on injection timing), have also strong influences. Further approaches to describe the burn rate have been published by List (triangular function, [121]), the double Wiebe function by Oberg [122], a polygon hyperbola function from Schreiner [123], or, as black-box, with neural networks [62].

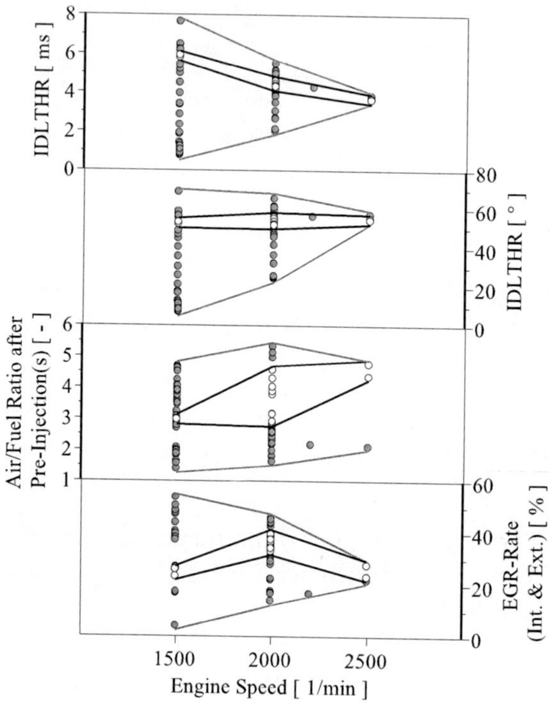

Figure 5.5: PCCI Measurements for the Empirical Based Model. The Gray Lines and Dots are Thresholds of the Measured Variations, the Black Line and White Dots Represent the Best Points for 1500, 2000 and 2500 min^{-1} and Load Variation, Respectively.

In order to apply the obtained results for the ignition delay, a model for the prediction of the ROHR could be further established. For this purpose, however, it is useful to define a constraint on the variability of some parameters.

From **Figure 5.5**, it can be derived that the best points only allow significantly restricted parameters. Therefore, the model would not have to be mapped for all parameters, but only for the spectrum, which can also be used with advantages in the application.

Figure 5.5 compares the ignition delay of low and high-temperature ignition, the air-fuel ratio after pre-injection, and the corresponding EGR rate for three different engine speeds. The gray lines notionally include all measured points (gray). The black lines only include the white points. These are the best points, which are discussed in section 4.5. The following conclusions can be drawn from this:

- The usable EGR rate becomes significantly lower with increasing engine speed. Extrapolation above 2500 min^{-1} shows that no EGR rate can be reasonably used. This was also demonstrated on the test bench.
- The usable bandwidth of the EGR rate and the air-fuel ratio after pre-injection is only narrow for the achievable best points. This results from the enclosed area of the black lines.
- The ignition delay exhibits a large spread at 1500 min^{-1}, which becomes significantly smaller with increasing speeds. Since the crank angle is plotted here as the axis scaling, the time-dependent ignition delay has an even more significant difference from 1500 to 2500 min^{-1}.

As the spreads of best points (covered in black lines with white points) show, the most satisfying application results can be achieved at 2000 min^{-1}.

This highlights the problem of PCCI. Good results are achieved, but the system can only be used advantageously in a limited operating range.

6 Summary and Conclusions

At the University of Stuttgart, the single-cylinder engine was investigated in terms of PCCI and conventional diesel application. PCCI was analyzed and evaluated regarding thermodynamic properties as well as raw engine emissions, indicating both advantages and disadvantages of this alternative combustion strategy. These results were used to improve the ongoing test bench measurements and subsequently to fit the three-stage Arrhenius model to predict ID for the start of low and high-temperature combustion. The comparison of PCCI with conventional diesel strategy led to the conclusion that the nearly homogeneous air-fuel mixture at SOC resulted in lower soot and NOx emissions at low to mid load (IMEP of 4–10) and engine speed (1500–2500 min^{-1}), respectively. On the other hand, the state-of-the-art piston geometry was used, and the comparison with the conventional diesel strategy shows disadvantages in CO and THC emissions and, consequently, in fuel consumption. This yields the following key findings:

- The advantage of simultaneous reduction of soot and nitrogen oxide was associated with the consumption disadvantage regarding the engine with a series application (piston, injector).
- Control of the combustion phasing by controlling the start of self-ignition via the main injection was possible to a limited extent. PCCI can therefore be regarded as a combination of HCCI and conventional diesel combustion.
- Improved results were achieved by ending the main injection before the start of the main combustion, thus decoupling injection and combustion. Therefore, the main injection did not burn diffusively, forming very low soot emissions. However, combustion phasing with decoupling of injection and combustion can only be determined by estimating the ignition delay.
- Several properties cause an increase in fuel consumption, such as the partial lean out of the mixture after pre-injections up to the chamber wall. Thus, this lean mixture does not reach combustible temperature, or wall

M. S. Wahl, *Emission Reduction with an Alternative Diesel Combustion Process*, Wissenschaftliche Reihe Fahrzeugtechnik Universität Stuttgart,
https://doi.org/10.1007/978-3-658-42094-9_6

interaction occurs before SOC. In addition, the injector geometry has disadvantages for pre-injections. As a result, there is the possibility of wall wetting in the case of unfavorable injection trajectories.

- In contrast to homogeneous premixed combustion (HCCI), which does not use an injection around TDC, partially premixed combustion with a main injection provides the option of controlling the combustion phasing [124]. However, using PCCI in transient conditions may not be sufficient to achieve benefits with state-of-the-art mechanical parts such as an omega piston.
- The penetration depth of the pre-injections has proven to be one main issue. However, investigations on engines with larger bores using nearly the same injection strategy as mentioned could lead to usable applications.
- Piston cooling provided a longer ignition delay, which allowed combustion phasing.
- Activation of EKAS for closing the swirl flap was not investigated. Literature shows that the effects counteract each other, and a positive influence on PCCI is not detectable. Homogenization leads to local lean areas and the ignition limit is exceeded. The pre-injected fuel remains unburned or reaches ignition limits after the main injection.

The three-stage Arrhenius model is suitable for predicting the low and high-temperature combustion ID. Subsequent work could focus on the ID of the main injection, whereby the physical part of the ID is more relevant than the chemical ID. Also, the hyperparameter optimization's computational effort could be used with an algorithm such as Rprop+ [119] to reduce the calculation time. First, however, a sensible threshold is relevant to find a termination criterion for the algorithm. This threshold could be the exceeding of the "coefficient of determination".

Nevertheless, the result of the multidimensional three-stage Arrhenius equation with 18 parameters will almost certainly provide more than one local minimum. The Rprop+ is intended to find a minimum next to the starting point. A combination of hyperparameter optimization, Rprop+, and a global optimization tool such as the H-Function described by Liu could be used to find a global minimum [125]. However, for all optimization tools, the exponential term is time-consuming and must be calculated for every degree of crank angle.

Coupling a zero- or quasi-dimensional combustion rate model with the presented ID model can increase predictability and accuracy for LT- and HTHR SOC. In addition, this ID model works well for short and long-time dependent delays with injection timing from SOI 30 to 100 °CAbTDC and a wide variation of test parameters.

Compared to previous concepts [126], the raw emissions of the PCCI injection strategy developed here are low, and further progress can lead to reduced fuel consumption. Using fuel with higher or lower CN in a diesel engine with the PCCI concept does not only affect the ignition delay time. Furthermore, the injection strategy must be adapted for fuels such as dimethyl ether (DME), Rapeseed Oil Methyl Ester (RME), or Methanol. However, satisfactory results for a reduction in engine raw emissions are possible. [21] Furthermore, these fuels can reduce soot with advanced evaporation properties and a lower tendency to form soot due to the fuel composition. Variable intake cam adjustment can be useful to control combustion based on the start of ignition and the combustion phasing. Early or late intake closing can reduce the effective compression well below the geometric compression ratio. However, the dynamic adjusting of all influencing parameters to achieve optimal combustion with MFB50 at around 8 °CAaTDC is complex since a misinterpretation of a signal for control of combustion (e.g., intake temperature or EGR valve position) can lead to very early and hard combustion.

PCCI operates based on the CN and the resulting long ignition delay time in the application shown. With increasing CN, the ignition delay is shortened to the measurements presented here. Changing CN implies that the combustion has to be actively adapted to the conditions such as temperature, fuel pressure, intake pressure, and cooling-water temperature by varying the injection using a closed-loop control system. However, a disadvantage is the limited operating range of PCCI, which might be further reduced by changing the fuel quality. On the other hand, a reduction in the CN would theoretically extend the ignition delay time and even widen the possible operating range. In order to use fuel deviating from the DIN EN 590, further investigations of the ignition limits and combustion must be carried out.

References

[1] M. Wahl, S. Schneider, M. Bargende (2022) An Empirical Based Model to Predict Ignition Delays in Partially Premixed Compression Ignition Mode. In: Bargende M, Reuss H-C, Wagner A (eds) 22. Internationales Stuttgarter Symposium. Springer Fachmedien Wiesbaden, Wiesbaden, pp 615–628

[2] M. Wahl, M. Bargende, A. Casal Kulzer et al. A Three-Stage Arrhenius Approach with a Coupled 0D-Model to Predict Ignition Delays in a Premixed Charge Compression Ignition (PCCI) Diesel Engine. In: Thiesel 2022 Conference on Thermo- and Fluid Dynamics of Clean Propulsion Powerplants

[3] R. Diesel (1893) Theorie und Konstruktion eines rationellen Wärmemotors zum Ersatz der Dampfmaschinen und der heute bekannten Verbrennungsmotoren. Julius Springer, Berlin

[4] A. Schmidt (2019) Technical Thermodynamics for Engineers. Springer International Publishing

[5] Diesel R. Internal Combustion Engine: Patent (608845)

[6] Diesel R. Vorrichtung zum Anlassen von Viertakt-Verbrennungskraftmaschinen durch Umwandlung derselben in Zweitakt-Druckluftmaschinen: Patent (86633)

[7] R. Diesel (1913) Die Entstehung des Dieselmotors. Springer, Berlin

[8] Statistische Ämter des Bundes und der Länder. Gemeinsames Statistikportal. https://www.statistikportal.de/de. Accessed 12 Nov 2022

[9] en2x Jährliche Verbraucherpreise Für Mineralölprodukte 1950-2021 (Inkl. Mwst) [€-Cent/Liter]. Accessed 12 Nov 2022 https://en2x.de/service/statistiken/verbraucherpreise/

[10] Kraftfahrtbundesamt Bestand nach Fahrzeugalter: Zeitreihen. https://www.kba.de/DE/Home/home_node.html. Accessed 12 Nov 2022

[11] Roland Berger (2015) Diesel controversy -Temporary shock or paradigm shift in powertrain? Impact of the diesel controversy on OEMs and suppliers.

M. S. Wahl, *Emission Reduction with an Alternative Diesel Combustion Process*, Wissenschaftliche Reihe Fahrzeugtechnik Universität Stuttgart, https://doi.org/10.1007/978-3-658-42094-9

[12] Die sparsamsten Autos. Spritmonitor.de: Verbrauchswerte real erfahren. https://www.spritmonitor.de/de/die_sparsamsten_autos.html. Accessed 12 Oct 2022

[13] Frontier Economics Ltd. (2020) The Concept of Efficiency in the German Climate Policy Debate on Road Transport: Translation from German original

[14] J. Demuynck, A. Kufferath, O. Kastner et al. (2020) Verbesserung der Luftqualität und des Klimas durch moderne Dieselfahrzeuge. MTZ Motortech Z 81:56–63.

[15] DIN e.V. (2022) Kraftstoffe – Dieselkraftstoff – Anforderungen und Prüfverfahren: Deutsche Fassung EN 590:2022 (590:2022). Accessed 2022-06

[16] EPA United States Environmental Protection Agency (2022) About Diesel Fuels. https://www.epa.gov/diesel-fuel-standards/about-diesel-fuels. Accessed 30 Jun 2022

[17] Mazda Motors Deutschland GmbH Mazda CX-60 Pressemappe. https://de.mazda-press.com/api/assets/download/428e5266-1726-4fee-b8ed-bb9c10abf585_Default?isDownload=false. Accessed 04 Aug 2022

[18] Tilles J. What are Squish and Swirl and Why Are They Important in a Diesel Engine? https://official.bankspower.com/what-are-squish-and-swirl-and-why-are-they-important-in-a-diesel-engine/. Accessed 2022-14-11

[19] J. B. Heywood (1988) Internal Combustion Engine Fundamentals. Automotive Technology Series. McGraw-Hill Book Co

[20] P. G. Eastwood, T. Morris, K. Tufail et al. (2007) The effects of fuel-injection schedules on emissions of NO x and smoke in a diesel engine during partial-premix combustion. In: SAE Technical Paper Series. SAE International400 Commonwealth Drive, Warrendale, PA, United States

[21] K. H. Mollenhauer (2007) Handbuch Dieselmotoren: Mit 580 zum Teil farbigen Abbildungen und 84 Tabellen, 3., neubearb. Aufl. VDI. Springer, Berlin, Heidelberg, New York

[22] W. Jaffé Pareto Translated: A Review Article. In: Journal of Economic Literature, 10, no. 4, pp 1190–1201

[23] P. Eckert, S. Rakowski (2019) Dieselverbrennung. In: Merker GP, Teichmann R (eds) Grundlagen Verbrennungsmotoren. Springer Fachmedien Wiesbaden, Wiesbaden, pp 639–665

[24] M. S. Gern (2021) Experimentelle Entwicklung einer Einspritzstrategie für Ottomotoren mit Wassereinspritzung. Springer Fachmedien Wiesbaden, Wiesbaden

[25] A. Vacca (2021) Potential of Water Injection for Gasoline Engines by Means of a 3D-CFD Virtual Test Bench. Springer Fachmedien Wiesbaden, Wiesbaden

[26] R. Plöntzke (2018) Einfluss der Wassereinspritzung auf den dieselmotorischen Verbrennungsprozess. In: Roß T, Heine A (eds) Der Verbrennungsmotor - ein Antrieb mit Vergangenheit und Zukunft. Springer Fachmedien Wiesbaden, Wiesbaden, pp 347–362

[27] J. Dec (1997) A Conceptual Model of DI Diesel Combustion Based on Laser-Sheet Imaging*. In: SAE Technical Paper Series, vol 970873. SAE International400 Commonwealth Drive, Warrendale, PA, United States

[28] P. F. Flynn, R. P. Durrett, G. L. Hunter et al. (1999) Diesel Combustion: An Integrated View Combining Laser Diagnostics, Chemical Kinetics, And Empirical Validation. In: SAE Technical Paper Series. SAE International400 Commonwealth Drive, Warrendale, PA, United States

[29] ECOpoint Inc. (2020) Diesel Fuel Injection: Multiple Injection Events. https://dieselnet.com/tech/diesel_fi.php. Accessed 26 Oct 2022

[30] A. G. Charalambides (2013) Homogenous Charge Compression Ignition (HCCI) Engines. In: Ng HK (ed) Advances in Internal Combustion Engines and Fuel Technologies. InTech

[31] G. Figer, G. Pirker, A. Wimmer et al. Analyse von HCCI-Brennverfahren beim direkteinspritzenden Dieselmotor: Analysis of HCCI Combustion Process on DI Diesel Engine. In: 9. Tagung "Der Arbeitsprozess des Verbrennungsmotors", vol 83, pp 275–290

[32] S. Kimura, O. Aoki, H. Ogawa et al. (1999) New Combustion Concept for Ultra-Clean and High-Efficiency Small DI Diesel Engines. In: SAE Technical Paper Series. SAE International400 Commonwealth Drive, Warrendale, PA, United States

[33] Y. Takeda, N. Keiichi, N. Keiichi (1996) Emission Characteristics of Premixed Lean Diesel Combustion with Extremely Early Staged Fuel

Injection. In: SAE Technical Paper Series. SAE International400 Commonwealth Drive, Warrendale, PA, United States

[34] K. Kumano, N. Iida (2004) Analysis of the Effect of Charge Inhomogeneity on HCCI Combustion by Chemiluminescence Measurement. In: SAE Technical Paper Series. SAE International400 Commonwealth Drive, Warrendale, PA, United States

[35] U. Egüz, N. Leermakers, B. Somers et al. (2014) Modeling of PCCI combustion with FGM tabulated chemistry. Fuel 118:91–99

[36] C. Krüger, F. Otto (2019) Simulation der Dieselverbrennung. In: Merker GP, Teichmann R (eds) Grundlagen Verbrennungsmotoren. Springer Fachmedien Wiesbaden, Wiesbaden, pp 1317–1334

[37] A. Kulzer, W. Fischer, R. Karrelmeyer et al. (2009) Kontrollierte Selbstzündung beim Ottomotor CO2 Einsparpotenziale. MTZ Motortech Z 70:50–57

[38] R. H. Stanglmaier, C. E. Roberts (1999) Homogeneous Charge Compression Ignition (HCCI): Benefits, Compromises, and Future Engine Applications. In: SAE Technical Paper Series. SAE International400 Commonwealth Drive, Warrendale, PA, United States

[39] A. W. Gray, T. W. Ryan (1997) Homogeneous Charge Compression Ignition (HCCI) of Diesel Fuel. In: SAE Technical Paper Series. SAE International400 Commonwealth Drive, Warrendale, PA, United States

[40] California Air Resources Board Airborne Toxic Control Measure for Diesel Engines on Commercial Harbor Craft Operated Within California Waters and 24 Nautical Miles of the California Baseline: Proposed 15-Day Modified Regulation Order: Proposed 15-Day Modified Regulation Order. Accessed 20th August 2022 https://ww2.arb.ca.gov/sites/default/files/barcu/regact/2007/chc07/reg17chc.pdf

[41] T. K. Sharma, G. A. P. Rao, K. M. Murthy (2016) Homogeneous Charge Compression Ignition (HCCI) Engines: A Review. Arch Computat Methods Eng 23:623–657

[42] S. Kook, C. Bae, P. C. Miles et al. (2005) The Influence of Charge Dilution and Injection Timing on Low-Temperature Diesel Combustion and Emissions. In: SAE Technical Paper Series. SAE International400 Commonwealth Drive, Warrendale, PA, United States

[43] D. L. Siebers, B. Higgins, L. Pickett (2002) Flame Lift-Off on Direct-Injection Diesel Fuel Jets: Oxygen Concentration Effects. In: SAE Technical Paper Series. SAE International400 Commonwealth Drive, Warrendale, PA, United States

[44] H. Ogawa, T. Li, N. Miyamoto (2007) Characteristics of low temperature and low oxygen diesel combustion with ultra-high exhaust gas recirculation. International Journal of Engine Research 8:365–378

[45] T. Kamimoto, M. Bae (1988) High Combustion Temperature for the Reduction of Particulate in Diesel Engines. In: SAE Technical Paper Series. SAE International400 Commonwealth Drive, Warrendale, PA, United States

[46] K. Akihama, Y. Takatori, K. Inagaki et al. (2001) Mechanism of the Smokeless Rich Diesel Combustion by Reducing Temperature. In: SAE Technical Paper Series. SAE International400 Commonwealth Drive, Warrendale, PA, United States

[47] H. Bockhorn (1994) A Short Introduction to the Problem - Structure of the Following Parts. In: Goldanskii VI, Schäfer FP, Toennies JP et al. (eds) Soot Formation in Combustion, vol 59. Springer Berlin Heidelberg, Berlin, Heidelberg, pp 3–7

[48] U. Wagner, S. Merkel, U. Spicher Untersuchungen zur Entwicklung einer rußfreien Verbrennung bei Dieselmotoren mit Direkteinspritzung: Institut für Kolbenmaschinen Universität Karlsruhe (TH). Förderkennzeichen: BWI 23001. In: Programm Lebensgrundlage Umwelt und ihre Sicherung (BWPLUS)

[49] Bockhorn H (ed) (1994) Soot Formation in Combustion: Mechanisms and Models. Springer Series in Chemical Physics, Vol 59. Springer Berlin Heidelberg, Berlin, Heidelberg

[50] A. Pungs Rußoxidation I + II: Untersuchungen zur Rußoxidation im Dieselmotor. In: FVV-Vorhaben, 526 + 607

[51] J. Nagle, R. F. Strickland-Constable (1962) Oxidation of Carbon Between 1000–2000°C. In: Proceedings of the Fifth Conference on Carbon. Elsevier, pp 154–164

[52] J. R. Walls, R. F. Strickland-Constable (1964) Oxidation of carbon between 1000–2400°C. Carbon 1:333–338

[53] M. Bargende (2010) Grundlagen der Verbrennungsmotoren: Vorlesung 3. Auflage, University of Stuttgart

[54] K.-H. Dietsche, K. Reif, et.al. (2018) Kraftfahrtechnisches Taschenbuch, 29th edn. Springer Vieweg

[55] Gutt E. Chemische Prozesse bei der Entstehung von Stickoxiden (NOx) in Diesel- und Benizinmotoren

[56] W. R. Dietrich (1999) Die Gemischbildung von Gas- und Dieselmotoren sowie ihr Einfluss auf die Schadstoffemissionen - Rückblick und Ausblick Teil 2. MTZ Motortech Z 60:126–134

[57] D. Rether (2012) Modell zur Vorhersage der Brennrate bei homogener und teilhomogener Dieselverbrennung. Dissertation, Universität Stuttgart

[58] W. Jost (2013) Explosions- und Verbrennungsvorgänge in Gasen. Springer Berlin Heidelberg, Berlin

[59] G. Sitkei (1964) Kraftstoffaufbereitung und Verbrennung bei Dieselmotoren. Springer Berlin Heidelberg, Berlin, Heidelberg

[60] Pischinger F. (2001) Abschlußbericht. Sonderforschungsbereich 224 "Motorische Verbrennung". http://www.sfb224.rwth-aachen.de/bericht.htm. Accessed 11 Nov 2022

[61] C. Barba (2001) Erarbeitung von Verbrennungskennwerten aus Indizierdaten zur verbesserten Prognose und rechnerischen Simulation des Verbrennungsablaufes bei Pkw-DE-Dieselmotoren mit Common-Rail-Einspritzung, ETH Zurich

[62] S. A. Beck (2012) Beschreibung des Zündverzuges von dieselähnlichen Kraftstoffen im HCCI-Betrieb. Dissertation, Universität Stuttgart

[63] Tschöke H (ed) (2007) Diesel- und Benzindirekteinspritzung: Einspritzsysteme - Potenziale, Anwendungen, Zukunftsentwicklungen. Haus-der Technik-Fachbuch. Expert-Verl., Renningen-Malmsheim

[64] F. Cupo, M. Chiodi, M. Bargende et al. (2019) Modelling of real fuels for an effective virtual engine development with focus on alternative fuel. In: Bargende M, Reuss H-C, Wagner A et al. (eds) 19. Internationales Stuttgarter Symposium. Springer Fachmedien Wiesbaden, Wiesbaden, pp 45–63

[65] S. Arrhenius (1889) Über die Dissociationswärme und den Einfluss der Temperatur auf den Dissociationsgrad der Elektrolyte. Zeitschrift für Physikalische Chemie 4U:96–116

[66] K. J. Laidler (1984) The development of the Arrhenius equation. J Chem Educ 61:494

[67] T. Nagy, T. Turányi (2011) Uncertainty of Arrhenius parameters. Int J Chem Kinet 43:359–378

[68] G. I. Senum, R. T. Yang (1977) Rational approximations of the integral of the Arrhenius function. Journal of Thermal Analysis 11:445–447

[69] J. Abraham, F. V. Bracco (1993) Simple Modeling of Autoignition in Diesel Engines for 3-D Computations. In: SAE Technical Paper Series. SAE International400 Commonwealth Drive, Warrendale, PA, United States

[70] H. Wolfer Der Zündverzug im Dieselmotor. In: VDI-Forschungsheft 392, Zündverzug und Klopfen im Motor, 9 Ausgabe B

[71] S. G. Bauer Ignition Lag in Compression Ignition Engines. In: Engineering, 148. July to December, 1939, pp 368–369

[72] M. A. Elliott (1906) Combustion of Diesel Fuel. In: SAE Technical Paper Series. SAE International400 Commonwealth Drive, Warrendale, PA, United States

[73] G. A. Weisser (2001) Modelling of combustion and nitric oxide formation for medium-speed DI diesel engines: a comparative evaluation of zero- and three-dimensional approaches. Dissertation, ETH Zurich

[74] A. Böge (2014) Handbuch Maschinenbau: Grundlagen und Anwendungen der Maschinenbau-Technik, 22., überarb. u. erw. Aufl. 2014. Springer Vieweg, Wiesbaden

[75] Bargende M, Wolgast EJ, Thiemann W (eds) (2011) Dieselmotorentechnik: Eine Wirkungsgradabschätzung moderner Diesel-Brennverfahren

[76] M. Bargende Berechnung und Analyse innermotorischer Vorgänge: Vorlesungsmanuskript SS 2017

[77] F. Weberbauer, M. Rauscher, A. Kulzer et al. (2005) Allgemein gültige Verlustteilung für neue Brennverfahren. MTZ Motortech Z 66:120–124

[78] Deutsche Normen (1976) Hubkolbenmotoren: Begriffe, Formelzeichen, Einheiten: Reciprocating internal combustion engines; terms, formulae, units (DIN 1940; ISO 2710-1973)

[79] Merker GP, Teichmann R (eds) (2019) Grundlagen Verbrennungsmotoren. Springer Fachmedien Wiesbaden, Wiesbaden

[80] Wankel F. (1957) Rotary Piston Machine (US2988008). Accessed 13 Nov 2022

[81] J. C. Kent, A. Mikulec, L. Rimal et al. (1989) Observations on the Effects of Intake-Generated Swirl and Tumble on Combustion Duration. In: SAE Technical Paper Series. SAE International400 Commonwealth Drive, Warrendale, PA, United States

[82] A. Floch, J. van Frank, A. Ahmed (1995) Comparison of the Effects of Intake-Generated Swirl and Tumble on Turbulence Characteristics in a 4-Valve Engine. In: SAE Technical Paper Series. SAE International400 Commonwealth Drive, Warrendale, PA, United States

[83] (2013) Valve Train: Systems and components, with 287 illustrations. MAHLE knowledge base, vol 3. Vieweg+Teubner Verlag, Stuttgart

[84] D. Bolaños Papenfort (2021) Experimentelle Ermittlung der Durchflussbeiwerte eines Einzylinder-Forschungsaggregates zur Anwendung in einer 0D-/1D-Simulationsrechnung: Studienarbeit, Universität Stuttgart

[85] D. Oh, C. H. Lee Flow coefficient measurements for an engine cylinder head under transient flow conditions with continuous valve lift change. In: International Journal of Engineering and Technology (IJET), Vol. 7 No. 3, pp 1135–1142

[86] S. Espig (2013) Beitrag zur Entwicklung eines Einspritzsystems für ein teilhomogenes Brennverfahren in Pkw-Dieselmotoren. Dissertation, Universität Stuttgart

[87] D. de Serio, A. de Oliveira, J. R. Sodré (2017) Effects of EGR rate on performance and emissions of a diesel power generator fueled by B7. J Braz Soc Mech Sci Eng 39:1919–1927

[88] J. Brettschneider Berechnung des Luftverhältnisses λ von Luft-Kraftstoff-Gemischen und des Einflusses von Meßfehlern auf λ Einflusses von Messfehlern auf λ. In: Bosch Technische Berichte, pp 177–186

[89] AVL (2005) Smoke Value Measurement with the Filter-Paper-Method: Application Notes. Accessed 17 Oct 2022 https://www.avl.com/documents/10138/885893/Application+Notes

[90] AVL List GmbH, Graz - Austria (2011, October) AVL 415S Smoke Meter GOO2 (From S/N 2500). Accessed 18 Oct 2022 https://www.avl.com/documents/10138/885893/Maintenance+Guide+415S

[91] International Organization for Standardisation (1998) Internal combustion compression-ignition engines - Measurement apparatus for smoke

from engines operating under steady-state conditions - Filter-type smokemeter (ISO 10054)
[92] A. E. Dodd, Z. Holubecki (195) The measurement of diesel exhaust smoke. Motor Industry Research Association (MIRA)
[93] A. C. Alkidas (1984) Relationships Between Smoke Measurements and Particulate Measurements. In: SAE Technical Paper Series. SAE International400 Commonwealth Drive, Warrendale, PA, United States
[94] G. G. Muntean (1999) A Theoretical Model for the Correlation of Smoke Number to Dry Particulate Concentration in Diesel Exhaust. In: SAE Technical Paper Series. SAE International400 Commonwealth Drive, Warrendale, PA, United States
[95] R. Christian, F. Knopf, A. Jaschek et al. Eine neue Messmethodik der Bosch-Zahl mit erhöhter Empfindlichkeit. MTZ 54 (1993) 1
[96] W. F. Northrop, S. V. Bohac, J.-Y. Chin et al. (2011) Comparison of Filter Smoke Number and Elemental Carbon Mass from Partially Premixed Low Temperature Combustion in a Direct-Injection Diesel Engine. Journal of Engineering for Gas Turbines and Power 133
[97] P. Kožuch (2004) Ein phänomenologisches Modell zur kombinierten Stickoxid- und Rußberechnung bei direkteinspritzenden Dieselmotoren
[98] S. Pöttker (2005) Optische Untersuchungen zum Einfluss synthetischer Kraftstoffe auf die homogene Kompressionszündung (HCCI). Dissertation, Hannover
[99] R. Miller, G. Davis, G. Lavoie et al. (1998) A Super-Extended Zel'dovich Mechanism for Nox Modeling and Engine Calibration. In: SAE Technical Paper Series. SAE International400 Commonwealth Drive, Warrendale, PA, United States
[100] Bundesministerium für Umwelt, Naturschutz und Reaktorsicherheit (1999) Nach "Bleifrei" kommt "Schwefelfrei": Bundesregierung beschließt neue Kraftstoffverordnung. Pressemitteilung 146/99. https://www.bmuv.de/pressemitteilung/nach-bleifrei-kommt-schwefel-frei
[101] Ohashi H., Saito J. (1995) Motor Exhaust Gas Analyzer Horiba MEXA-7000 Series: 1. Product Concept. Accessed 15 Nov 2022 https://www.horiba.com/uploads/media/R011-04-019-E.pdf.
[102] Takeda K., Koike H. (1995) Motor Exhaust Gas Analyzer Horiba MEXA-7000 Series:2. Downsizing and Modular Configuration of Ana-

lyzers. https://www.horiba.com/uploads/media/R011-05-025-E.pdf Accessed 15 Nov 2022

[103] (2022) Three-way catalyst (TWC): Operation of a three-way catalyst. https://ac.umicore.com/en/technologies/three-way-catalyst/

[104] Diesel Oxidation Catalyst (DOC). Accessed 15 Nov 2022 https://ac.umicore.com/en/technologies/diesel-oxidation-catalyst/

[105] P. G. Burman, F. DeLuca (1962) Fuel Injection and Controls: For Internal Combustion Engines. Library of Congress Catalog Card Number: 62-12020. The Technical Press Ltd., London

[106] C. Weiskirch, W. Gotre, B. Röthlein et al. (2017) The development of the common rail fuel injection system – A brief view in the mirror on ups and downs. In: Bargende M, Reuss H-C, Wiedemann J (eds) 17. Internationales Stuttgarter Symposium. Springer Fachmedien Wiesbaden, Wiesbaden, pp 1347–1369

[107] Walter M., Palmer J., Rupp A. et al. Method and Device for Zero-Fuel Quantity Calibration of a Fuel Injector NECTOR (US 2015/0034.048 A1)

[108] J. Sauter Die Grössenbestimmung der im Gemischnebel von Verbrennungskraftmaschinen vorhandenen Brennstoffteilchen. In: VDI-Forschungsheft, vol 279

[109] S. Ashley Can Diesel Finally Come Clean? In: Scientific American, a Division of Springer Nature America, Inc.

[110] H. Tschöke, K. Mollenhauer, R. Maier (2018) Handbuch Dieselmotoren. Springer Fachmedien Wiesbaden, Wiesbaden

[111] K. Reif (op. 2010) Moderne Diesel-Einspritzsysteme: Common Rail und Einzelzylindersysteme: mit 133 Abbildungen, 1. Aufl. Bosch-Fachinformation Automobil. Vieweg + Teubner, Wiesbaden

[112] Kistler Gruppe (2020) ThermoCOMP Quarz Pressure Sensor: Cylinder pressure sensor for combustion engines. Type 6041A… 6041C_003-591e-11.21. https://www.kistler.com/files/document/000-013e.pdf Accessed 28 Oct 2022

[113] T. D. Pedersen, J. Schramm (2010) Reduction of HCCI Combustion Noise Through Piston Crown Design. In: SAE Technical Paper Series. SAE International400 Commonwealth Drive, Warrendale, PA, United States

[114] K. Finger (2001) Untersuchungen zur Kraftanregung durch die Verbrennung beim direkteinspritzenden Common-Rail Dieselmotor unter Berücksichtigung des Körperschallübertragungsverhaltens. Dissertation, Technische Universität

[115] P. Eilts Zur Ablagerungsbildung im schwachlastbetrieb hochaufgeladener Dieselmotoren. In: VDI-Forschungsheft, 57 (1991) No. 665, pp 1–44

[116] C. Barba, C. Burkhardt, K. Boulouchos et al. (1999) Empirisches Modell zur Vorausberechnung des Brennverlaufes bei Common-Rail-Dieselmotoren. MTZ Motortech Z 60:262–270

[117] C. Barba, C. Burkhardt, K. Boulouchos et al. (2000) A Phenomenological Combustion Model for Heat Release Rate Prediction in High-Speed DI Diesel Engines with Common Rail Injection. In: SAE Technical Paper Series. SAE International400 Commonwealth Drive, Warrendale, PA, United States

[118] G. Stiesch, G. P. Merker (1999) A Phenomenological Model for Accurate and Time Efficient Prediction of Heat Release and Exhaust Emissions in Direct-Injection Diesel Engines. In: SAE Technical Paper Series. SAE International400 Commonwealth Drive, Warrendale, PA, United States

[119] C. Igel, M. Hüsken (2003) Empirical evaluation of the improved Rprop learning algorithms. Neurocomputing 50:105–123

[120] I. I. Vibe (1970) Brennverlauf und Kreisprozeß von verbrennungsmotoren: Übersetzung aus dem Russischen. VEB Verlag Technik Berlin

[121] H. List (1939) Thermodynamik der Verbrennungskraftmaschine. Springer Vienna, Vienna

[122] H. J. Oberg (1976) Die Darstellung des Brennverlaufes eines mittelschnelllaufenden Dieselmotors durch zwei überlagerte Vibe-Funktionen: Dissertation

[123] K. Schreiner Untersuchungen zum Ersatzbrennverlauf und Wärmeübergang bei Hochleistungsdieselmotoren. In: Motortechnische Zeitschrift MTZ, vol 54, pp 554–563

[124] E. Shim, H. Park, C. Bae (2020) Comparisons of advanced combustion technologies (HCCI, PCCI, and dual-fuel PCCI) on engine performance and emission characteristics in a heavy-duty diesel engine. Fuel 262:116436

[125] X. Liu (2001) Finding Global Minima with a Computable Filled Function. Journal of Global Optimization 19:151–161

[126] H. Yokota, Y. Kudo, H. Nakajima et al. (1997) A New Concept for Low Emission Diesel Combustion. In: SAE Technical Paper Series. SAE International400 Commonwealth Drive, Warrendale, PA, United States

Appendix

A1. Diesel Fuel Test Result

Diesel Analysis at 03-11-2021 with number FB21-00257 – KRE based on DIN EN 590:

- Cetane Number DIN EN 17155
- Density DIN EN ISO 12185
- Polycyclic aromatic hydrocarbons DIN EN 12916
- Sulfur content DIN EN ISO 20844 / 20846
- Hydrogen content DIN EN ISO 12937
- C/H ratio DIN 51732
- Oxygen Content DIN 51732
- Lower and upper heating value DIN 51900-1/-2

Table see next page.

M. S. Wahl, *Emission Reduction with an Alternative Diesel Combustion Process*, Wissenschaftliche Reihe Fahrzeugtechnik Universität Stuttgart,
https://doi.org/10.1007/978-3-658-42094-9

Table A1.1: Test Bench Configuration.

Name	Measurement Parameter	Test method	Unit	Test result	Limit Value (Norm) DIN EN 590:2014	
					Min.	Max.
Diesel red	Cetane Number	DIN EN 17155:2018	-	51.9	51.0	-
	Density (15 °C)	DIN EN ISO 12185:1997	kg/m^3	829.1	820	845
	Monoaromatic Hydrocarbons	DIN EN 12916:2019	% (m/m)	18.5	-	-
	Diaromatic Hydrocarbons			2.1	-	-
	Triaromatic Hydrocarbons			0.4	-	-
	Polyaromatic Hydrocarbons			2.6	-	8.0
	Total aromatics			21.1	-	-
	Sulfur content	DIN EN ISO 20884:2019	mg/kg	8.8	-	10
	Water content	DIN EN ISO 12937:2002	mg/kg	< 30	-	200
Diesel red	Carbon content	DIN 51732:2014	% (m/m)	86.0	-	-
	Hydrogen content			13.8	-	-
	Oxygen content	DIN 51732:200 mod.	% (m/m)	< 0.5	-	-
	Gross calorific value (Ho,v)	51900-1:2000 mod.	J/g	45926	-	-
	Calorific value (Hu,p)	DIN 51900-2:2003 mod.	J/g	42994	-	-